Renewable Energy: A Very Short Introduction

VERY SHORT INTRODUCTIONS are for anyone wanting a stimulating and accessible way into a new subject. They are written by experts, and have been translated into more than 45 different languages.

The series began in 1995, and now covers a wide variety of topics in every discipline. The VSI library currently contains over 600 volumes—a Very Short Introduction to everything from Psychology and Philosophy of Science to American History and Relativity—and continues to grow in every subject area.

Very Short Introductions available now:

Available soon:

For more information visit our website

www.oup.com/vsi/

Nick Jelley

RENEWABLE ENERGY

A Very Short Introduction

OXFORD
UNIVERSITY PRESS

OXFORD
UNIVERSITY PRESS

Great Clarendon Street, Oxford, OX2 6DP,
United Kingdom

Oxford University Press is a department of the University of Oxford.
It furthers the University's objective of excellence in research, scholarship,
and education by publishing worldwide. Oxford is a registered trade mark of
Oxford University Press in the UK and in certain other countries

First edition published in 2020

Impression: 6

Published in the United States of America by Oxford University Press
198 Madison Avenue, New York, NY 10016, United States of America

British Library Cataloguing in Publication Data

Data available

Library of Congress Control Number: 2020930603

ISBN 978-0-19-882540-1

Printed in Great Britain by
Ashford Colour Press Ltd, Gosport, Hampshire

Contents

Acknowledgements

I am very grateful to the following for discussions about renewable energy: John Andrews, Conyers Davis and colleagues at the Schwarzenegger Institute in Los Angeles, Nick Eyre, Chris Goodall, Mike Mason, Moritz Riede, Gerard van Bussel, and Simon Watson. Also I would to thank Latha Menon for her advice, and all the staff at OUP who helped prepare the manuscript for publication.

Above all, I would like to thank my wife Jane whose support and comments were invaluable, and John, Tessa, and Piernicola for their encouragement.

List of illustrations

List of abbreviations

AC	alternating current
AI	artificial intelligence
CHP	combined heat and power
CO_2	carbon dioxide
DC	direct current
eV	electron-volt, unit of energy
GDP	gross domestic product
$GtCO_2$	gigatonne (thousand million tonnes) of CO_2
GW	gigawatt, thousand MW
GWp	gigawatt peak, thousand million Wp
HVAC	high voltage AC
HVDC	high voltage DC
kW	kilowatt (thousand watts), unit of power
kWh	kilowatt-hour, unit of energy
kWp	kilowatt peak, thousand Wp
LED	light emitting diode
MW	megawatt, thousand kilowatts
MWh	megawatt-hour, thousand kWh
PV	photovoltaics
TW	terawatt, thousand GW
TWh	terawatt-hour, thousand million (billion) kWh
Wp	Watt peak, unit of power of a solar module

Chapter 1
What are renewables?

We use energy all the time in our daily life: when we make a call
on our mobile, boil a kettle, or drive a car. Energy is vital for a
good quality of life: in providing warmth, in producing food,
and in powering technology; and in the last 200 years, we have
increasingly relied on fossil fuels. However, burning coal, oil, and
natural gas to supply energy pumps huge amounts of carbon
dioxide (CO_2) into the atmosphere, and also produces harmful
pollutants that damage our health and the environment. It would
be easy to carry on as we are, and there are enough deposits of
fossil fuels to last for several hundred years. But the level of carbon
dioxide in the atmosphere, which has already started to seriously
disrupt our climate, would cause dangerous climate change before
the end of this century because of global warming, and put many
millions of lives at risk. And right now, the air pollution is already
causing seven million premature deaths every year.

Renewable energy

Fortunately, some of the energy we use does not have these
damaging consequences, in particular energy generated by solar,
wind, and hydro power. And in many parts of the world, solar and
wind generators are becoming the cheapest source of power, and a
viable alternative to fossil fuels. Moreover, these energy sources

are renewable, as they are naturally replenished within days to decades. When the energy produced is affordable, and its generation is not damaging to the environment or to people (as can happen, for example, when forests are cut down for bioenergy plantations), then the supply of renewable energy is sustainable. Such sustainable energy could provide many immediate benefits. It could reduce the air pollution that plagues many of our cities today, provide cheaper energy and many new jobs, and give energy security to millions at an affordable cost.

Only a few years ago, giving up our reliance on fossil fuels to tackle global warming would have been very difficult, as they are so enmeshed in our society and any alternative was very expensive. Also, because climate change appeared to be only a gradual and distant threat, which did not prompt an emotional response and immediate action, many individuals and governments were reluctant to act. But now the threat is much closer, and moving away from fossil fuels to renewables has become essential.

The Sun is our main source of renewable energy. Its radiation provides light for illumination, heat for warmth and cooking, and can also be converted by photovoltaic cells into electricity. Variations in its absorption across the Earth's surface generate differences in surface temperatures that cause winds, and these can be used to drive wind turbines. Rivers are our source for hydropower; and are part of the natural cycle where rainfall from clouds flows via rivers into the sea, and the evaporation of water into the atmosphere through solar heating results in clouds. Plants derive their energy through photosynthesis, in which carbon dioxide and water are converted to carbohydrate in the presence of sunlight, and are used to provide food, and also fuels for heating and engines. The wind and solar resources each have the potential to provide all the energy that the world needs, solar many times over.

Early sources of energy

Nearly all of the sources of energy up to the 18th century were from renewables. Plants and animals provided food, and materials such as wood, dung, oil, and fat, for cooking, heating, lighting, and shelter; and these are referred to now as traditional biomass. By the Bronze Age, charcoal, which is produced by burning wood in a reduced supply of air, and which burns at a higher temperature, enabled metals to be extracted from ores. Animals made many tasks easier, with horses and oxen providing power and transport for many centuries.

The main use of solar energy, besides its importance for growing food, was for heating and lighting in dwellings. The ancient Greeks and Chinese built houses with the main rooms facing the south to catch the Sun, and the Romans also added glass windows, which helped retain the heat. In hot dry climates, houses tended to be built with very thick walls to keep out the heat during the day and to provide warmth at night. The heat from the Sun was also used for drying materials, such as clay for bricks, and for preserving foods.

Some coal, peat, and crude oil were used to provide heat and light, though their contribution was small, and tended to be confined to regions where this fuel was easily obtained. In a few places, geothermal energy was available in the form of hot springs. This source of energy is renewable as it comes from heat given out from the interior of the Earth. The Romans utilized these springs in some of their bath houses, as can be seen in Pompeii.

Wind power

Records exist of sailing boats on the Nile 5,000 years ago. Early sailing ships had a single mast with a square sail, and were propelled along by their sail when there was a following wind.

But these ships could also travel obliquely into the wind by angling their sail to just catch the wind. The sail curved, and the wind created a pressure difference across it that pushed the sail in a direction perpendicular to that of the wind—just as the air flowing over an aircraft wing gives lift. Early Mediterranean galleons, such as the Roman trireme, combined oars and sails, as did the Viking ships; but most merchant vessels were sailing ships. Triangular sails, called lateen sails, that evolved as a simpler and cheaper alternative to square rigged sails, became well established by the 5th century, and also allowed ships to tack into the wind (Figure 1).

Sailing ships opened up sea trade all over the world, and evolved into multi-mast vessels with both square and lateen sails, which could travel great distances at some speed. The tea clippers in the mid-19th century could travel at over 15 knots (28 kilometres per hour), and average about half that speed; similar to the speeds of modern racing yachts. Where possible, moving goods by sea was generally much cheaper and easier than by land, as roads were often very rudimentary.

The first recorded use of wind power on land was in the 10th century in Persia, where vertical axis windmills were used to pump water and grind grain. Similar windmills were used in China and may have developed there first. These machines were less efficient than the horizontal axis windmills that first appeared in Europe in England, France, and Holland around the 12th century. Such windmills spread rapidly eastward in Europe in the

1. **Lateen sails.**

13th century, often initially in places without access to watermills. They were used mainly for grinding grain, pumping water, and sawing wood. The first mills were mounted on posts and could be manually turned to face the oncoming wind. In later and larger mills, which were introduced around the 14th century, only the top with the sails and windshaft swivelled. The use of large windmills peaked in the 19th century, by which time coal-fired steam engines were taking over, which were more compact and available on demand.

Hydropower

Waterwheels were developed in Egypt, China, and Greece over 2,000 years ago, in the two centuries before the first millennium. They were first used for irrigation and supplying drinking water, and by the beginning of the first millennium they had been modified to operate machines such as millstones and saws. A Roman factory with sixteen waterwheels for milling grain operated in Gaul in the 2nd century, and watermills were widespread in the Chinese and Roman Empires by the 5th century. Their use for industrial processes had spread to Islamic countries by the 10th century, and to Europe by the 12th century where, for example, the Cistercian monks used them for forging metals and making olive oil.

In the waterwheels used for raising water, the water flowed underneath a vertically mounted wheel. Buckets attached to the rim of the wheel were filled and emptied as the wheel turned, and it is thought that some Arabian wheels could lift water 30 metres. More power could be achieved by flowing the water across the top of a wheel, filling compartments on its rim. The movement and weight of water turned the wheel. Horizontal waterwheels, which had a vertical shaft and water flowing across one side of the wheel, were simpler to construct and well suited to fast flowing streams. The motion of tides was also used on a small scale in medieval times to power watermills on the coast. Tidal energy is also renewable, because the losses arising from tidal motion and in taking energy from the tides have only a negligible effect on the relative motion of the Moon, Earth, and Sun.

The industrial revolution was first powered by water. Richard Arkwright employed waterwheels in 1771 to drive his cotton spinning frames in the first textile mill, in Derbyshire, England. The technology was significantly improved in 1827 by the French engineer Benoît Fourneyron. He enclosed the water flow down a wide cylindrical tube, and past fixed curved blades. These directed the water horizontally onto all the moving blades, which were attached to a vertical shaft (see Figure 2). The turbine was compact, and allowed the flow and pressure of water (and hence the power) to be much higher. It was also very efficient (over 80 per cent) and, with a small fall (head) of water of 1.4 m and a 2.9 m diameter turbine, his machine at Fraisans in France in 1832 produced around 37 kilowatts of power. It was a major advance and the forerunner of modern water turbines.

While coal-fired steam engines replaced water power for cotton spinning in England, water-powered turbines remained important in France and the USA during the 19th century. At the end of the 19th century electricity generation using huge water turbines was developed. The first large hydroelectric plant in the USA, built by

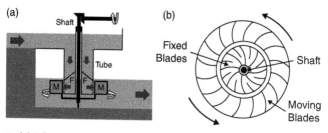

2. (a) Schematic of a Fourneyron turbine: the fixed blades F direct the water flow through slots in the tube onto all the moving blades M, which are attached to a vertical shaft—see (b) Horizontal section through the turbine wheel housing.

Nikola Tesla and George Westinghouse in 1895 on the Niagara Falls, employed three Fourneyron turbines each capable of delivering 3,700 kilowatts.

The rise of fossil fuels

In Britain, the use of coal increased steadily after the 12th century due to the shortage of fuel in the form of wood and charcoal. Pollution from burning coal, a problem we are familiar with today, led to unsuccessful attempts to ban its use in London from the 13th century onward. Coal use really took off in the industrial revolution in the 18th century, with coal displacing charcoal in the smelting of iron; and demand was further fuelled by the development of powerful steam engines by Watt and Bolton in the latter half of the 18th century. The industrial revolution spread to Europe, North America, and then worldwide, and trade expanded massively with the introduction of coal-fired steamships and steam trains, which greatly increased the global demand for coal.

Crude oil had been used in relatively small quantities for millennia in China, Arabia, Central Asia, and elsewhere. Exploration for oil for lamps, and for use in lubrication, was already happening in North America and Europe in the mid-19th century, but the start of the modern oil industry is generally credited to Edwin Drake, who devised a method of protecting the sides of a hole drilled in gravel from collapsing, by installing a tube down to the bedrock below. In 1859 in Pennsylvania in the USA he struck oil at a depth of 21 m and had a yield of 20–40 barrels a day. Unfortunately, he did not patent his invention and failed to make money. Demand grew with the development of internal combustion engine driven cars in the late 19th century, and cars and trucks are now the main consumers of oil, refined into petrol and diesel.

Gas was first produced commercially by the gasification of coal in the early 19th century, and initially provided street lighting. The introduction of electric lighting towards the end of the 19th century led to coal gas being used for cooking and heating. Natural gas from oil fields was also employed, and by the latter part of the 20th century had completely displaced coal gas. Natural gas is now increasingly being burned in power plants for the generation of electricity in place of the more polluting coal.

The resurgence of renewables

Other than hydropower, which grew steadily during the 20th century and now provides almost a sixth of the world's electricity demand, renewable energy was a neglected resource for power production for most of this period, being economically uncompetitive. But the oil price crises of the 1970s caused Western governments to begin sponsoring research programmes into various renewable energy technologies, with the aim of reducing their dependence on oil. Wind power was the first technology to become commercially viable, benefiting from low capital costs and tax breaks, and using the knowledge of blade design in the aircraft industry. However, the development of wind power in the 1970s and 1980s waxed and waned with the price of oil. Despite this, some nations with access to strong winds and concerns over the availability of affordable energy (energy security) maintained support for it. There was also a gradually increasing awareness of the dangers from global warming. Now wind power is cost competitive with fossil-fuel generation in many regions.

Wave power also attracted a lot of interest in the 1970s, but it was soon recognized that the capital costs were high and that most devices were unable to withstand severe storms at sea. Nonetheless, a few designs are still under development, particularly ones which are submerged and tethered to the sea floor. Tidal power is economically viable only in those regions with a large tidal range or where the average tidal flow is sufficiently fast. This will limit its contribution, but a few areas are suitable: some of the best being off the coasts of North America, and around the UK.

Photovoltaic cells took much longer to gain a foothold in the energy market. In the photovoltaic effect a voltage is generated when certain materials are illuminated, and this phenomenon was originally observed in 1839 by Edmond Becquerel. But it was not until the 1950s that Bell Laboratories pioneered the development of silicon photovoltaic cells, with efficiencies of around 6 per cent. However, their high cost meant that they were restricted to niche applications in satellites and the space programme. The oil crises of the 1970s spurred interest in photovoltaic cells, and over the last few decades mass production methods have reduced the cost of solar cells enormously. We are now at the point where electricity from solar farms is commercially competitive in many parts of the world.

On a different front, the growth of the biofuels industry in the last part of the 20th century has slowed in the past decade, with concern over associated carbon dioxide emissions from land clearances and conflicts with food production. However, biomass is an important source of energy in the developing world. And, as long as emissions in planting and harvesting are negligible, it is carbon-neutral, in that the carbon dioxide produced by burning the material is reabsorbed by new crops.

Policies supporting renewables

Renewable generators, other than hydropower, were initially more expensive than fossil-fuel plants, and required subsidies in order to compete, and one of the most successful mechanisms has been feed-in tariffs. These are a guaranteed price that a producer of renewable electricity will receive; the price is set so that a reasonable profit can be made, and the long-term period of the tariff reduces the risk for investors. The extra cost of production over that for fossil-fuel generation is usually shared by all the consumers of electricity in the province or country. Germany,

Denmark, Spain, and the United States pioneered the creation of markets for renewables, which led to advances in technology and economies of scale. But manufacture and installation of renewables is now dominated by China.

The global increase in production has led to the cost of electricity generated from renewables approaching that from fossil fuels, and auctions are now an increasingly popular way of promoting renewable energy deployment. In these auctions, generator companies submit a bid with a price per kilowatt-hour (kWh) for a certain amount of electricity. The competitive auctions bring out the real price of generation, often difficult for a regulator to determine, and revenue for the successful company is guaranteed for a period, which is good for investment. These auctions have led to significant cost reductions over the last few years.

The importance of energy in society

The energy supplied by fossil fuels enabled the industrial revolution to expand first in Britain and then throughout the world. What was behind such a great transformation of society was the flexible and significant power (available on demand) that initially coal-fired, and later internal combustion engines, could produce. A standard measure of power is the kilowatt (kW) and is roughly the power a single horse can produce steadily over a long time. It corresponds to lifting a weight of 50 kg at a speed of 2 metres per second. Increasing the speed to 4 metres per second would require 2 kW; so, using more power enables tasks to be completed more quickly.

Early waterwheels of a couple of metres diameter could produce a few kilowatts of power, as could 17th-century Dutch windmills; but the development of efficient and powerful steam engines by James Watt and Matthew Boulton at the end of the 18th century paved the way for steam power. The early engines generated about 5 kW, but the use of high-pressure steam, introduced by Richard

Trevithick in 1805 in the first steam train, enabled much more powerful and lighter engines to be produced. Engines with over 50 kW power were available by 1850 and by then trains could travel at 80 kilometres per hour (50 mph). Steam trains and steam boats opened up trade across the world. Steam engines also allowed factories and towns to expand in any region without concern over the availability of water power, and their use grew throughout the mid 19th century. The turn of the century saw the emergence of electricity and the internal combustion engine as important features of industrialized societies.

The greatly increased access to power boosted manufacturing output and trade, and over time saw a general rise in the standard of living, with wages increasing and more than doubling in the UK from 1840 to 1910. Technological improvements had led to increased productivity, but the rapid change initially caused hardship for the many required to work long hours in difficult and sometimes dangerous conditions. In England life expectancy only started to rise after 1860, when public health measures were introduced. However, by the end of the 19th century, the industrial revolution, powered by fossil fuels, gave much improved living conditions in many countries: heating, lighting, and labour saving machines in homes, better public facilities, and much faster transport, among many other things. It follows that access to energy is very important for a good quality of life.

A simple measure of the standard of living in a country is the Human Development Index (see Figure 3), which combines indicators of educational attainment, life expectancy, and income. While it does not include any measure of inequality, it does emphasize that development is not just about economic growth. A good correlation is found for low values of the Human Development Index of a country with its energy use. However, a large spread in energy consumption per person is seen between different highly developed countries. Although some of this reflects differences in climate, considerable scope exists for

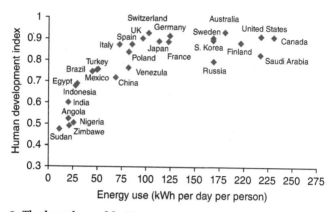

3. The dependence of the Human Development Index on the average energy use in 2013. The average energy use is the average power consumption in a country, which is mainly in buildings, industry, and transport, divided by its population; e.g. an average of 2 kW per person is equivalent to an energy use of 48 kWh (kilowatt-hour) per day per person.

reducing consumption by improvements in efficiency and changes in lifestyle.

Less-developed countries will seek to improve their standard of living by increasing their average energy use. In particular, just under one billion people (13 per cent of the global population) were still without electricity in 2018. Even a small amount of electricity per person can enable access to mobile phones, computers, the web, lighting, TV, and refrigeration. So, the cost of energy is an important factor in the economy of a society, and for many decades, this meant that fossil fuels were often the preferred choice.

Global energy usage

The annual global demand for energy is an enormous number of kWh, and can be more easily grasped in terms of terawatt-hours (TWh), where a TWh equals 1,000 million (a billion) kWh. The

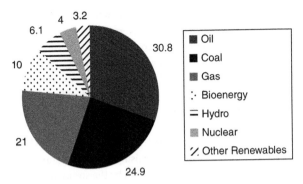

4. Total energy consumption in 2017 showing the percentage contribution of different sources.

global demand in 1800 was about 6,000 TWh for around 1 billion people; and was nearly all provided by traditional biomass. In 2017, it was twenty-five times more (155,000 TWh) for 7.6 billion people. Figure 4 shows the percentages of the global total energy consumption in 2017 provided by the main energy sources, almost 80 per cent of which were fossil fuels. Other renewables include wind, solar, and geothermal power, with the fastest growth seen in wind and solar farms. Bioenergy is mainly energy produced from traditional biomass.

About a third of the total energy consumption is used mainly in the conversion of fossil fuels to electricity and refined fuels. The rest is called the final energy demand and is the energy consumed by users: about 100,000 TWh per year. About 10 per cent is as heat from traditional biomass in the developing world, 22 per cent as electricity, 38 per cent as heat predominantly from fossil fuels, and 30 per cent in transport. Both the heat and electricity are mostly used in industry and buildings. Petrol (gasoline) and diesel provide nearly all of the fuel used for transport.

We see that providing heat is just as important as electricity. Both can be measured in kWh, but while electricity can be fully changed into heat, as in an electric oven, only a fraction of the energy in the form of heat can be converted into electricity, as some is necessarily lost to the surroundings. In a thermal steam power plant, the chemical energy bound up in a fossil fuel is transformed as the fuel is burnt into thermal energy. This heats water to generate steam, which expands across turbine blades that turn an electrical generator. Only part of the heat is transformed into electricity; the rest is transferred to the environment as waste heat when the steam is condensed, completing the cycle. The fraction converted is increased by raising the temperature of the high-pressure steam, but this is limited by the strength of the boiler tubes at high temperature. A typical percentage of heat converted to electricity in a modern thermal steam power station is about 40 per cent. For a combined cycle gas turbine (CCGT) plant, which operates at a higher temperature, the percentage can be as large as 60 per cent.

Similarly, only a fraction of the heat generated in an internal combustion engine can be transformed into the energy of motion (kinetic energy) of a vehicle; a typical average efficiency is 25 per cent for a petrol and 30 per cent for a diesel fuelled car, while for diesel trucks and buses the efficiency can be about 40 per cent. Electric motors, on the other hand, have efficiencies of around 90 per cent, so electrifying transport would reduce the consumption of energy significantly. This is an example of the synergy between improved efficiency and renewable energy, which will help in providing the energy that the world needs.

The renewable resource of hydropower helped initiate the growth of electrical grids at the end of the 19th century, and in 2018 met 16 per cent of the world's electricity generation. Investment in the other renewables—wind, solar, geothermal, and bioenergy—occurred much later, in the last few decades of the 20th century. Growth at first was slow, since these renewables were not cost competitive and subsidies were required. But as production increased, costs came down and their contribution started to increase. The percentage of electricity generation from these other renewables has now risen to 9.7 per cent in 2018, up from 3.5 per cent in 2010, and, including hydropower, the total contribution of renewables is 26 per cent.

But as a percentage of the global energy, rather than just electricity, consumed by users, renewables only provide about 18 per cent, with about 10 per cent supplied by traditional biomass. Their share though is now poised to increase significantly in the coming decades, as solar and wind power become cheaper than fossil-fuel generation in many countries. But it has taken a long time for the world to realize that renewables must be the dominant source of energy from now on.

Chapter 2
Why do we need renewables?

We have relied on fossil fuels for over two centuries, but it is only in the last thirty years or so that the dangers from burning fossil fuels have really started to be appreciated. Most countries now accept that we need to replace fossil fuels, but it is very difficult to do so, because they are so convenient, widely available, and, until recently, generally the cheapest source of energy. Coal and natural gas are used extensively for heating in power plants, industries, and homes; and oil and its derivatives are compact sources of energy, which are well suited for powering cars, trucks, ships, and planes. Moreover, because the fossil-fuel industry and infrastructure are so well established, strong vested interests exist in maintaining our dependence on them.

But the use of fossil fuels causes global warming and climate change. This is now leading to widespread concern, and also to a growing realization of the harm caused by the air pollution from coal burning, and from internal combustion engines in cars and lorries. These threats are causing a switch away from fossil fuels to renewables that is gaining impetus from the growing awareness of the increased intensity and frequency of extreme weather seen in recent years. This transition is also being aided by the falling price of clean energy from renewables. But why has it taken so long to realize the dangers of relying on fossil fuels?

The dangers of fossil fuels

Coal-fired stations powered the industrial revolution first in England, and then around the world—most recently in China and India, where they have contributed significantly to serious pollution in cities, with dust and smog particles reducing visibility and affecting health. Burning coal also emits sulphur dioxide, which causes acid rain, damaging buildings and vegetation; and coal mining can cause serious environmental degradation. Conventional oil is increasingly inaccessible, and deep-water drilling has raised the risk of major oil spills, such as occurred at the Deepwater Horizon platform in 2010. While extracting gas and oil from shale deposits has become economically viable in the last decade, and shale oil and shale gas have made a significant contribution to production, their impact on the environment is causing concern—notably the pollution of groundwater from fracking. But it is the emission of CO_2 that is the greatest danger.

As long ago as 1896 the Swedish scientist Svante Arrhenius pointed out that the carbon dioxide from burning fossil fuels could raise global temperatures. But many thought the amount emitted would be absorbed by the oceans and land, and would not have any significant effect. Concern was raised again in 1938 when Guy Callendar argued that warming was already occurring, but this conclusion was contested. However, measurements in 1960 by Charles Keeling showed that the levels of CO_2 in the atmosphere were clearly rising, and his experiments were sufficiently accurate that the effect on the growing season could be seen.

The picture was confused by the cooling caused by the dust and smog particles in the atmosphere from increased industrial activity after the Second World War. However, by 1979 a US National Academy of Science panel concluded that a doubling of the pre-industrial level of CO_2 would raise global temperatures by about 3 °C. This deduction has been reinforced since then by increasingly sophisticated climate modelling. The reason that this temperature rise occurs is because CO_2 is what is called a greenhouse gas.

The greenhouse effect and global warming

Sunshine, which is composed of visible light, infrared, and ultraviolet radiation, passes through our atmosphere, and maintains the Earth's temperature at the point where it loses heat at the same rate as it receives it. The heat loss into space is by the emission of infrared radiation, and its rate is governed by the temperature near the top of the atmosphere. Matching this heat loss to the heat gained from the Sun fixes the temperature at the height in the atmosphere where the air is thin enough that the infrared radiation escapes. The Earth's surface is warmer than at this height, as the lower atmosphere acts like a blanket and blocks most of the infrared radiation emitted by the Earth: the atmosphere heats up and in turn heats the Earth. The infrared radiation from the Earth, which has much longer wavelengths than the radiation from the Sun (since the Earth's temperature is much lower), is absorbed by the greenhouse gases in the atmosphere, which are mainly water vapour, carbon dioxide, and methane. The trapping of infrared radiation and the consequent temperature rise is called the greenhouse effect.

In 1824 the great French mathematician Jean Fourier was the first scientist to realize that the atmosphere acts like a greenhouse. However, it was not until 1859 that the Irish scientist John Tyndall identified the main gases in the atmosphere that absorb infrared radiation. In particular, he realized that we actually need the greenhouse effect, as otherwise 'the Sun would rise upon an island held fast in the iron grip of frost'. The principal greenhouse gas is carbon dioxide, and its amount, prior to the burning of fossil fuels, was maintained at an approximately constant level by natural processes, which include photosynthesis that removes CO_2, and respiration and decomposition that generate CO_2.

However, increasing the concentration of greenhouse gases by burning fossil fuels raises the height at which the air is thin enough for the infrared radiation to escape. The thickness of the absorbing lower atmosphere (the insulating blanket) therefore increases, and as a result the surface temperature of the Earth rises. This warming was commented upon in a newspaper article as long ago as 1912 (see Figure 5). While the estimate of the amount burnt was roughly one and a half times too large, it was a very prescient remark.

The average warming since pre-industrial times is now about 1 °C, and how the average global temperature has changed since the end of the 19th century (when accurate temperature measurements were available) is shown in Figure 6. While some

---

COAL CONSUMPTION AFFECT-ING CLIMATE.

The furnaces of the world are now burning about 2,000,000,000 tons of coal a year. When this is burned, uniting with oxygen, it adds about 7,000,000,000 tons of carbon dioxide to the atmosphere yearly. This tends to make the air a more effective blanket for the earth and to raise its temperature. The effect may be considerable in a few centuries.

---

5. *The Rodney and Otamatea Times*, Waitemata and Kaipara Gazette, 14 August 1912.

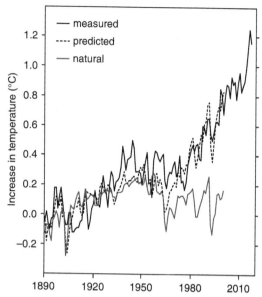

6. Curves showing global temperatures 1890–2018: measured; predicted (natural plus human induced changes); and estimated natural changes.

of the observed temperature changes are caused by variations in the intensity of solar radiation, and by natural phenomena such as volcanic activity, the rise over the period 1970–2000 (and up to 2018) can only be explained by the human induced change in the atmospheric concentration of greenhouse gases, in particular CO_2 from fossil-fuel burning. The temperature of the Earth would have been expected to remain about the same as in the 1970s in the period 1970–2000 (and up to 2018), if only natural causes were included.

The measured temperatures for the period 2003–13 were roughly constant, and were used by some to cast doubt on whether global warming was actually occurring. But over the longer period 1990–2018 the observations are consistent with an overall steady rise in temperature, as predicted from an increasing level of greenhouse gases. Fluctuations caused by natural causes, such as an El Niño or a La Niña, which is a warming or cooling of the surface of the sea in the Pacific that occurs every few years, will cause deviations that can mask a steady rise in temperature.

Climate change

The emission of greenhouse gases, in particular CO_2, mainly from the combustion of fossil fuels, but also from land use changes (such as replacing forests with crops), causes not only global warming but other changes. The melting of glaciers and of the Greenland and Antarctic ice sheets contributes to sea-level rises, and the loss of ice is affecting animal habitats. At the same time, acidification of the oceans, through more dissolved CO_2, threatens shellfish and other sea creatures, and is causing large amounts of coral to die worldwide. Many species have changed their geographic locations due to changes in climate arising from global warming. A small rise in temperature can heighten the risk of droughts and can affect crop yields: even a 1 °C warming reduces the average wheat yield by 6 per cent.

In the Arctic, ice melting exposes more water, which is less reflective than ice, so more sunshine is absorbed, which amplifies the warming. The region is now over 2 °C hotter than in the 1970s and significant areas of the Arctic are ice-free as a result. This polar amplification increases the chance of blocking patterns of high pressure associated with extreme heat waves, such as the one experienced in Europe in the summer of 2018. It also brings an increased risk of changes in the jet stream, which can, for instance, cause cold Arctic air to move and bring very cold weather further south. Now that the atmosphere is generally hotter because of global warming, clouds can hold more water vapour. This means that when severe storms occur, rainfall can be much heavier and severe flooding is more likely. Snowfalls too can be much deeper. So even one degree of global warming can seriously disrupt our climate and significantly increase the frequency of extreme weather events.

The effects of global warming of greater than 2 °C could be catastrophic for many parts of the world. It could accelerate the melting of glaciers in the Andes and western China, threatening the water supply of millions of people as the seasonal stores that glaciers provide, which hold water in the winter and release it in the summer, are lost. The rise in sea levels would flood low-lying islands, such as the Maldives, and inundate coastlines, for example, Bangladesh and parts of the east coast of the USA, leading to massive movements of populations. There would also be an increased threat to human health from higher temperatures, particularly in tropical and subtropical regions. Biodiversity is also likely to be irreversibly altered, with the possible extinction of a significant percentage of plant and animal species. Ironically, some of the poorest nations, who have contributed least to climate change, will be most affected by it. The livelihoods of indigenous people and of many millions working the land are already affected by extreme droughts, flash flooding, and erratic seasons. And such extreme weather is precipitating migration.

If the world continues using energy produced mainly from fossil fuels, the danger is considerable that the temperature rise by 2100 will be too high to avoid devastating effects. As a result, there is an overwhelming consensus in the scientific community, and increasingly among politicians, that decisive action must be taken now to reduce our carbon emissions. In the Paris Agreement of 2015, nations agreed to limit global warming to 2 °C, and if possible move towards a limit of 1.5 °C. Although President Trump announced in 2017 that he intended to withdraw the USA from the agreement, many states in the USA, notably California, are still committed to lowering emissions.

A further report in 2018 by the International Panel on Climate Change found that the risks were significantly lower for a global warming of 1.5 °C than for 2 °C. Limiting the rise to 1.5 °C could spare hundreds of millions of people from climate-related disasters by 2050 and reduce the loss of biodiversity significantly. But while action is clearly vital and urgent for the long-term health of the world, as the warming is already 1 °C, limiting emissions has many immediate benefits for the quality of life of communities, and it is these, in particular, that are motivating people to expand renewables.

Carbon dioxide emissions must fall

Carbon dioxide concentrations in the atmosphere were about 280 parts per million (ppm) in 1750 (just before the industrial revolution), and rose slowly at first to about 310 ppm in 1950; but since then have increased rapidly, to 350 ppm in 1990 and 407 ppm in 2018. The lifetime of carbon dioxide added to the atmosphere is about 200 years before most of it is absorbed. The emissions of other non-CO_2 greenhouse gases, of which methane is the largest, make up about a quarter of the total, However, methane has a much shorter lifetime of about ten years, which means that, although it is a much more powerful greenhouse gas than carbon dioxide, it is the cumulative emissions of carbon dioxide that largely determine the global mean surface warming.

Since pre-industrial times, the amount emitted up to 2017 was about 2,200 thousand million (giga) tonnes of carbon dioxide, which has caused a temperature rise of approximately 1.1 °C. Allowing for non-CO_2 effects, we must limit further emissions of CO_2 to 580 gigatonnes to restrict global warming to about 1.5 °C. This means rapidly reducing our dependence on fossil fuels—if we carry on as we are (emitting about 37 gigatonnes of CO_2 a year from fossil fuels), we will have exceeded that temperature by 2035, only a short time away. (This is in line with the rise in temperature seen in Figure 6.)

Although fossil fuels were formed from biomass, the process of fossilization is far too slow to replenish the fossil fuels—we burn in one year what took around a million years to lay down. But we cannot rely on 'running out' of fossil fuels to make us stop using them; burning all the proven reserves of coal, oil, and gas would emit around 3,000 $GtCO_2$. In addition, there are large amounts of unconventional and undeveloped resources, such as shale oil, shale gas, and tar sands. So fossil fuels are limited by emissions not by resources: most must remain in the ground to keep global warming to less than 1.5 °C; they must become 'frozen assets'.

Renewables as alternatives to fossil fuels

We must take urgent action and we need non-CO_2 emitting sources of energy to replace coal and gas in the generation of both electricity and heat, and oil in transport. Replacements in the form of renewables must be as cheap as fossil fuels, as energy is vital for a good standard of living, and few people are in a position to afford anything other than the cheapest energy sources. We can do this because solar and wind already provide cost-competitive electricity in many regions, and ultimately either could meet the total global demand. As a result, solar and wind will be the dominant forms of renewable generation in the future. Hydropower is also as economic a renewable, though resources are much more limited, with its potential contribution around 15 per cent of the present global demand.

However, while all these renewables can now be as cheap as fossil fuels, they are not as compact a source of energy. Wind, solar photovoltaic, and hydropower installations all take up much more area than an equivalent fossil-fuel power plant. For many applications this is not a problem: solar panels can be mounted on roofs or in fields or deserts, wind turbines on hilltops, and water turbines in rivers; but for transport, in particular for cars, a compact source is essential. Since these renewables all produce electricity, this source can now be a lithium-ion battery, which is lightweight and compact. Fossil fuels provide a lot of heat, particularly in industry, and using renewables to provide heat, either directly, as in an electric-arc furnace, or indirectly to produce zero-carbon fuels, such as hydrogen, will increase the demand for electricity significantly.

Although biomass is potentially renewable and can be a source of heat for industry and in thermal power plants, the massive areas of land required to produce significant amounts of energy raises concerns, in particular, over competition with food production; and research is now concentrating on plants that can grow on land unsuitable for food crops.

Economics of renewable power plants

It is vital that renewable energy is economic and that depends both on the efficiency of the power plant and its capital cost, as well as on the interest rate on the capital, the lifetime of the plant, and on the cost of maintenance and operation. For example, a 20 per cent efficient silicon solar cell can produce electricity much more cheaply than a 40 per cent efficient multijunction cell, because the multijunction cell is currently a lot more expensive. While the size of a solar plant will be smaller if its efficiency is larger, it is the price per kWh that predominantly matters economically.

The maximum output of a power plant is generally called its capacity, or rated capacity. Capacity is expressed in megawatts, MW, where a MW is 1,000 kilowatts. For fossil-fuel power plants, typical capacities are around 1,000 MW, while large wind and solar photovoltaic (PV) farms are often now about 500 MW, and large hydropower plants several thousand MW. Power plants do not run all the time, such as when there is a lack of demand for electricity during the night, and in the case of renewable plants, a lack of resource, such as wind or sunshine. So, the amount generated in a year is a fraction of what the capacity would produce if it were running continuously, and this fraction is called the capacity factor. For a coal-fired power plant the capacity factor is about 0.6. As the speed of wind varies and the sun only shines during the day, the capacity factors are less for wind and solar farms: the average is currently around 0.3 for wind onshore and 0.5 for wind offshore; and around 0.10–0.25 for solar, depending on the location.

A 500 MW wind farm with a capacity factor of 0.4 would generate about 1,750 million kWh (1.75 TWh) of electricity a year, sufficient for half a million European homes. A rough estimate for the cost of electricity from a modern turbine with a lifetime of twenty years would be 5 eurocents per kWh, competitive with fossil-fuel generation. Better winds giving a capacity factor of 0.5 would reduce the cost to 4 eurocents per kWh, and increasing the lifetime of the turbine would also lower the cost. The cost of capital, which is dependent on the interest (or discount) rate, has a marked effect; for example, changing the discount rate from 8 per cent to 4 per cent would decrease the cost of electricity by 20 per cent. This rate depends on many factors, in particular the confidence in the project being built on time and to cost, as well as on the market interest rates.

Land or sea area required for renewables

Unlike fossil-fuel power plants, wind and solar farms require considerable areas to generate large amounts of power. Generally, enough space is available; and in the case of wind, the land between turbines can be used for grazing or crops. In the United States, wind farms covering only 2 per cent of the land area could supply the total electricity demand. Even where the population density is high, as in parts of Europe, wind can make a significant contribution out of sight by placing the wind turbines offshore. The wind resources in the North Sea are among the best in the world: a wind farm covering an area of about 30 square kilometres will produce a thousand million kWh (1 TWh) per year, enough electricity for about a quarter of a million homes in Europe.

The UK's total annual consumption of energy is 1,650 TWh, and to supply half of this would take about 3.5 per cent of the North Sea, an area about 20 per cent larger than that of Wales. Distribution hubs and undersea transmission lines would be required, but suitable areas exist in waters less than 50 m deep off the UK. However, when this estimate was made in 2009, offshore wind was thought prohibitively expensive. Since then, though, costs have been coming down very fast such that, by the early 2020s, offshore farms will start to be as cost competitive as new gas-fired power stations, and by the end of the 2020s as economical as existing ones.

When considering a developing country, it is not the present energy demand that matters, but what amount is required to give a good quality of life. In India, and even more so in Africa, many millions of people are without access to electricity. The energy demand per person for a good standard of living, as measured by the Human Development Index (see Figure 3), is around 80 kWh per day. But this demand includes the energy used in the conversion of fossil fuels to electricity, and the fossil-fuel energy consumed in transport. Electrifying transport and producing electricity directly with renewables will reduce this energy demand to about 60 kWh per day. For India and Africa, with their high solar intensity, photovoltaic farms could supply a significant fraction of this demand.

In India, the space required to generate 1 TWh per year is about 15 square kilometres. So, with a population of 1.35 billion and an area of 3.3 million square kilometres, about 7 per cent of the land would be needed to meet half the total energy demand; in Africa, the corresponding percentage is 1 per cent. (Non-arable land could be used to avoid competition with food production.) Africa therefore could potentially power itself entirely with solar photovoltaic panels; and even export electricity, which could generate wealth.

Variability of renewable energy

Unlike fossil-fuel generators, whose fuel can be stored ready for use, neither solar nor wind are always available. This is not a problem unique to renewables, as conventional generators can go offline when a fault arises. Back-up supplies are then needed, as they are when peaks in the demand for electricity occur; these are typically gas-fired generators. While wind and solar can complement each other, with wind speeds typically higher in the winter than in the summer, and sunshine greater in summer than in winter, there can still be significant gaps in availability.

As the fraction of electricity generated by renewables increases, this variability in supply becomes larger. This can be met by having more back-up generators, which will need to be without significant CO_2 emissions (low-carbon), or by using energy storage units, such as batteries. These units would store surplus generation that could be used when more electricity is required. The demand can also be altered (demand response) to meet the supply and can be smoothed, for example by moving electricity consumption from the evening to the daytime, using smart meters and controls. Interconnectors that make available generators located over a large region or number of countries can also help meet peaks in demand. Even when there is no wind or sun, the lights can be kept on.

Alternative low-carbon sources

While the emissions from fossil-fuel fired utility power stations are more than ten times those associated with the manufacture and operation of renewables, technologies are available that would reduce these significantly. The emitted CO_2 can be captured chemically and then pumped into underground stores, such as disused gas fields. The technology could also be applied to industrial processes. The capital costs of carbon capture, though, are high and there has not been the support to make it competitive. But interest in applying the technology to capturing CO_2 directly from the air is growing. This air capture could help to reduce the cumulative emissions of CO_2 should we fail to stop burning fossil fuels fast enough. But based on progress so far, it does not appear likely that carbon capture could provide more than a 10 per cent reduction in emissions by 2050.

Nuclear power plants generate electricity without carbon dioxide emissions; however, their uptake has been affected by concerns over their safety—following the very serious accidents at Chernobyl in Ukraine and Fukushima in Japan—and over their cost, which in some countries is now making nuclear power uncompetitive. There are also fears that the technology could be used to enrich uranium for nuclear weapons, and concerns over the disposal of reactor waste. While these have caused a move away from nuclear power in Europe and North America, there has been some expansion in Asia; by 2050, its contribution to the world's electricity requirement is expected to be relatively small at around 10 per cent.

So, while these alternative low-carbon technologies will help, they are not adequate to provide the power we need without any carbon dioxide emissions, which will need to happen by about 2050 to avoid dangerous climate change. We need to see how much the 'traditional' renewables—biomass, solar heat, and hydropower—as well the more modern ones—wind (using turbines), and solar (using photovoltaic panels)—could each contribute by 2050, and at what cost; and why the other renewable technologies—tidal, wave, and geothermal—are only expected to make a small contribution. We will see that we can both displace fossil fuels with their damaging emissions, and meet the demand for energy that raising people's standard of living requires, but it will require an enormous worldwide effort to do it in time.

Chapter 3
Biomass, solar heat, and hydropower

Biomass

Since the earliest human settlements, we have grown plants and hunted animals for food and burnt wood for warmth. There is energy in this biomass: in the heat from burning it and in the food we and animals eat. This energy comes from the Sun, as plants grow using photosynthesis to capture sunlight to convert carbon dioxide from the air and water from the ground into carbohydrate. With the rising global population, land resources are being drastically depleted. A considerable amount of land is required to grow what we need, and using biomass for energy can clash with growing plants for food and with preserving ecosystems. Currently biomass provides about 10 per cent of all the energy we consume—primarily as heat from the burning of wood, charcoal, dung, or the residues from crops for cooking and heating homes—and provides about the same amount of energy as food. This traditional biomass is the main source of energy for many people in the developing world.

The oil shortages in the 1970s stimulated interest in obtaining fuels from crops as a substitute for the oil-derived fuels: petrol (gasoline) and diesel. Two of the main biofuels are ethanol, which can be produced by the fermentation of sugar from sugar-containing plants, such as corn; and biodiesel, which can be made from plant oils, such as palm oil. There has also been interest in biomass, such as wood and agricultural residues, that can be burnt, instead of coal or gas, either in power stations that generate electricity, or in industrial processes that require heat. These bioenergy crops are potentially a low-carbon sustainable source of energy, provided negligible amounts of CO_2 are released when planting and harvesting these crops.

The efficiency of conversion of solar energy to biomass in photosynthesis is low, at around 1 per cent. This means that large tracts of land are needed to grow bioenergy crops. For example, using all the arable land in Japan would only provide an amount of biofuel equal to 30 per cent of the volume of its gasoline consumption each year. So, finding suitable land for biomass can be a problem.

Traditional biomass

Traditional biomass provides energy for about 2.5 billion people in the developing world. Another 0.3 billion rely on coal and kerosene. However, the smoke from burning wood, charcoal, coal, or kerosene in simple stoves and open fires is very damaging, with about 3.8 million premature deaths each year, mostly affecting women and children. They are often the ones spending hours collecting wood: time that the children could be in school and the women could use on other activities.

About 85 per cent of the 1 billion people living in sub-Saharan Africa are dependent on traditional biomass. Much of it is used as charcoal, particularly in urban areas, because of its compactness and ease of use—it is made traditionally by heating wood in earth pits or mound kilns with a limited supply of air. Demand is expected to rise because of the growing population and because more people are living in cities. This has raised concerns over deforestation and land degradation. Unregulated harvesting of wood can have devastating environmental effects, as we can see in

7. **The Haiti Dominican Republic border showing the marked contrast between deforested (Haiti) and forested (Dominican Republic) land.**

the deforestation in Haiti compared with the preserved forest in the Dominican Republic (see Figure 7).

Attempts worldwide have been made to introduce improved cooking stoves, with China in the 1980s and early 1990s introducing 130 million. But most have only limited health benefits, as the burning of the fuel still produces particulates in the smoke and some carbon monoxide. To remove these harmful emissions requires complete combustion. This can be achieved by introducing heated air above the burning fuel, but this tends to make the stove complex and expensive and not many people in rural communities can afford one. Still, the introduction of simple credit schemes that use mobile phones is giving access to good stoves to some on lower incomes.

The cost of these clean biomass stoves needs to come down, and centralized industrial production would help through economies of scale. Because of the sharply falling cost of solar panels and of batteries, another solution may soon be electric stoves.

But many people are reluctant to use new cooking methods and are unaware of their health benefits. And in some poor neighbourhoods, the fear of theft can put off investment in new technology.

Biofuels

There has been interest in biofuels since the end of the 19th century, when Rudolf Diesel demonstrated his first engine at an exhibition in Paris using peanut oil; and later in the 1920s when Henry Ford envisioned tractors running on ethanol produced from crops that could be fermented. However, interest waned after the Second World War with the availability of cheap oil from the Middle East. It was renewed when oil supplies were threatened in the 1970s, and policies were enacted in many countries to encourage the growth of biofuels; initially to provide energy security, but then also to combat global warming.

One of the most successful biofuel programmes is in Brazil, where there are very large sugar cane plantations, and where ethanol production originally started in the late 1920s. The energy yield is high since relatively little energy is used in growing and harvesting the crop, and in producing the ethanol. Large areas of suitable pasture land and good weather conditions are available for growing sugar cane, and ethanol now contributes about a third of the fuel used by cars in Brazil. However, other biofuel programmes have been less successful. In the USA, farmers have been given incentives to make ethanol from corn (maize), but the energy yield is low, since a lot of energy is required to produce the corn. Moreover, the amount of bioethanol produced in the USA is only about 10 per cent of the volume of its gasoline consumption. To produce another 10 per cent would require around 15 per cent of the whole US cropland.

In Europe, biodiesel was encouraged by the European Commission and vegetable oil production rose, particularly in the first decade of this century. The oils have to be first treated chemically before they can be used in modern diesel engines, as they are thicker than diesel fuel. The demand for palm oil for biodiesel led to increased deforestation in Indonesia and Malaysia, and the draining of large areas of peatlands. This caused decomposition of the peat and led to fires that released huge amounts of CO_2, which will take many years to offset through the regrowth of biomass. Tropical deforestation has also been caused through clearing land for cattle, soybeans, and wood, and is responsible for about 10 per cent of greenhouse gas emissions worldwide. It has also contributed to a serious loss in biodiversity. Now the expansion of all biofuels has slowed, due to concern over their related CO_2 emissions, their impact on the environment, and their potential conflict with food production; for instance, both palm oil and corn are important foods.

Environmental impact and advanced biofuels

All these concerns over environmental impact have shifted the focus to advanced (also called second-generation) biofuels. These are fuels from plants that grow on areas unsuitable for food crops. Cellulose-based plants, such as switch grass, survive on poor land. But breaking up the cellulose with acids to yield sugar that can be fermented to ethanol is expensive. However, it was noticed in the Second World War that a fungus broke down cotton clothes and tents by secreting an enzyme. Little energy is required using enzymes in the production of ethanol; however, making this method cost effective has proved much harder than was anticipated.

Microalgae have also attracted a lot of interest as a possible answer, as they can have high oil content and can grow in salty or waste water on arid land. But after many years of research, producing biofuel from microalgae is still not commercially competitive. Generally high oil yield is at the expense of growth rate, and growing the microalgae and extracting the oil has proved expensive. Genetically modified strains are being considered, but, as microalgae play a very important role in regulating marine food chains, the concern is they would upset the delicate balance of species in the environment if they were released.

Anaerobic digestion of plants, food, and animal waste in purpose-built vats produces biogas (predominantly methane), which can be burnt for heating and cooking, or used in power plants (see Figure 8). This process also occurs naturally in compost heaps, and in the stomachs of cows and other ruminants; the latter is estimated to account for about 5 per cent of global warming. Small digestors are widely used in Asian villages; and in China, over 30 million homes use biogas. But its take-up in sub-Saharan Africa has been poor, often because it is too labour intensive, and increasingly solar power is meeting the demand for energy.

Biomass has an important role as a sustainable source of carbon in the manufacture of many products, displacing the use of oil. Plants have evolved that have adapted well to semi-arid conditions, such as cacti; their leaf pores remain closed during the day to reduce water loss, and only open at night to capture CO_2. Globally, roughly 25 million square kilometres exist of semi-arid

8. A one megawatt (1 MW) anaerobic digestion plant in Wales, turning food waste into electricity.

and idle crop lands, distributed widely in both the developing and developed world, and about 10 per cent of this area may be available for growing well-adapted plants. These could be an important source of biomass for the chemical industry that would avoid competition with food production, and the emissions arising from the use of oil.

Biomass potential

Currently the main use for modern biomass is as bioenergy crops for heating in buildings and industry and power generation, where it contributes about 4 per cent of global demand. This is where considerable potential lies, as the main sources are forestry, agricultural residues, and municipal wastes, which have less environmental risk than bioenergy crops. Large amounts of these resources are currently untapped. Typically to be competitive with fossil fuels, the supply of biomass needs to be readily available, as in Sweden where it provides almost a quarter of their energy supply. In the production of electricity, the biomass can be used very effectively when electricity generation is combined with the heating of buildings. As it can be available on demand, bioenergy is also a useful complement to the variable sources of wind and solar. But globally, the lack of economic competitiveness is limiting expansion. Also, the large areas of suitable and accessible land required are not widely available: 3,000 square kilometres, that is, a square with 55-kilometre sides, are needed for growing the biomass for a typical thermal power station of 1,000 MW output, which would power around 1.5 million households in Europe.

Biofuels, which were promoted as a substitute for oil-derived fuels, are only contributing around 3 per cent of the demand. Renewably generated electricity stored in batteries is looking increasingly likely to be the preferred low-carbon replacement in cars, rather than biofuels. However, if advanced biofuels can be produced economically and sustainably, they might make a significant contribution in aviation, as a substitute for jet fuel, and in shipping as a replacement for oil. Biomass could also be a sustainable source of carbon for the chemical industry.

Currently, the contribution of biomass, mostly traditional, is some 15,000 TWh per year. Concern over sustainable production means that by 2050 only a similar contribution might be made, mainly through modern biomass, as the growing access to electricity through solar panels will reduce the demand for traditional biomass significantly. Such a contribution, though, will need strong policy support and regulation to ensure that negligible CO_2 emissions are associated with planting and harvesting of crops, and to protect food production, biodiversity, and land rights.

Solar heat

The Sun has provided heat for buildings for millennia, but it was only at the end of the 19th century, when homes in the USA were getting mains water, that a commercial solar water heater was produced. These solar collectors consisted of a water tank painted black, enclosed in a glass fronted box. By connecting the collector to an insulated tank located above, hot water could be stored for use at night. During the day, water heated in the collector flowed up to the storage tank and colder water flowed down (see Figure 9). These thermo-syphon systems were popular in California, until gas was found in Los Angeles in the 1920s. They also flourished in Florida until losing out to electric water heaters after the Second World War. The interest in renewables in the 1970s revived interest in many countries, with Israel taking a lead in passing legislation to compulsorily install solar water heaters on new buildings.

Many modern homes now have solar thermal collectors on their roofs. Two types are prevalent: flat plate collectors, which are similar to the earlier collectors in design; and evacuated tube collectors, which reduce heat loss from convection by having a

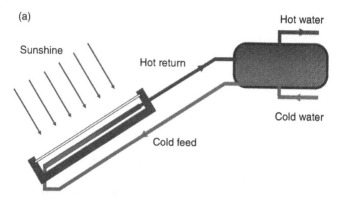

Hot water

Sunshine

Hot return

Cold water

Cold feed

9. (a) Thermosyphon system. (b) Solar water heaters mounted on rooftops in China.

transparent pipe, from which the air has been removed, enclosing the blackened water pipes. China now accounts for over two-thirds of the world market, and putting this technology and good insulation into all new buildings during construction would reduce significantly the carbon footprint of buildings, that is, the total carbon dioxide emitted per year. Solar heating is also increasingly being used in industrial applications and for district heating. But worldwide the total contribution is small at about 400 TWh in 2018.

Solar heat for generating power

Solar heat was first introduced commercially for producing power in 1913 in Egypt by an American engineer, Frank Schuman. He used five parabolic trough-shaped mirrors to concentrate sunlight onto pipes carrying water, to generate steam to operate a pumping engine that produced more than 40 kW. The British and German governments planned similar units to provide energy in their colonies. However, the need for energy in a readily transportable form during the First World War drove a rapid expansion in oil exploration, and concentrated solar power almost disappeared. The shortages of oil in the 1970s revived interest, and the first commercial plants, which used parabolic trough collectors, began operation in the Mojave Desert in California in the 1980s. In the 1990s, a solar tower system was built at Barstow in California and demonstrated how solar energy could be stored. A large array of mirrors was used to direct the Sun's rays onto a tank of molten salt on top of a central tower. The heated salt (at over 500 °C) was pumped into a large storage vessel, and then used to transfer heat to a boiler for a conventional thermal plant that produced 10 MW of electricity. When there was no sunshine, the hot molten salt in the storage vessel could be used as a source of heat for the thermal power plant for several hours (see Figure 10). However, the decrease in the cost of fossil fuels in the 1990s, and the withdrawal of incentives, stifled any significant growth until 2006, when government and state initiatives in Spain and the USA revitalized the market.

In the 2000s it was thought that concentrated solar power plants would be able to generate electricity more cheaply than

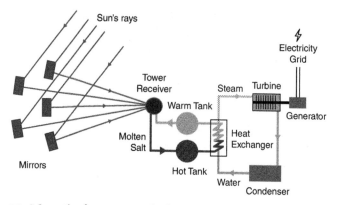

10. Schematic of a concentrated solar power plant. Molten salt is circulated to generate steam to run a turbine generator. When the sun is down, the molten salt in the hot tank (565 °C) is pumped through the steam generator to the warm tank (kept at 285 °C, which prevents the salt from solidifying).

photovoltaic systems. One of the plants designed at that time was the Crescent Dunes tower plant in Nevada, USA (see Figure 11), which has close to 10,000 mirrors that direct sunlight onto a receiver at the top of a 200-metre-tall tower. It was planned to give a maximum output of 110 MW and 10 hours of storage using molten salt, and was the first of its kind. But its high capital cost meant that the price of electricity was 13.5 cents per kWh, which was no longer competitive with solar photovoltaic farms, whose costs had fallen very sharply. (Moreover, it has had technical problems that have hurt its performance.)

So instead of some concentrated solar power plants being built, photovoltaic farms were constructed. Since then, efforts have been made to improve the efficiency of conversion of solar to electrical energy by operating at higher temperatures, while at the same time lowering the cost of components and of storage. In the USA, the Sunshot programme is aiming to reduce the cost of electricity to 6 cents per kWh by 2020.

11. Crescent Dunes 110 MW concentrated solar power plant.

Outlook for solar heat

Solar heating for homes and industry is facing competition from electrically driven systems, but might well generate 2,000 TWh per year by 2050. The outlook for concentrated solar power is encouraging, for despite the differences in costs and a reduction in feed-in tariffs, the availability of supply after sunset offered by concentrated solar power plants with thermal storage can be a significant advantage over solar photovoltaic farms. This is because storing electricity in batteries is currently (2018) more expensive than using molten salt to store the equivalent amount of heat. This advantage, which enables it to act as an effective back-up supply, gives it added value and has meant this technology still attracts considerable interest. And in 2019 in Morocco, a 150 MW solar tower power plant with storage exceeded performance targets on output and storage in the first few months of operation. China is investing heavily, and projects are planned or under construction in several other countries where there is good sunshine and clear skies, for instance South Africa, Chile, Kuwait, Israel, India, and Saudi Arabia. Their environmental impact, particularly when compared to fossil-fired plants, is relatively benign.

Research and development have brought down costs, and in 2018 there were contracts signed for power plants with storage to generate electricity at less than 7 US cents per kWh in Australia and in Dubai. If these operate as planned, then such power plants, as they can generate electricity on demand, could help significantly in integrating wind and solar photovoltaic farms in a grid, while maintaining a low cost of electricity. In 2010, the International Energy Agency thought that concentrated solar power plants could provide about 5,000 TWh a year by 2050 (10 per cent of the expected global electricity demand), and with the cost reductions seen since then, this looks reasonable.

Hydropower

In a hydropower plant, the energy in water falling from a height is converted to electrical energy using a turbine (see Figure 12(a)). Many large installations use a type of turbine designed by James Francis in 1848. In this turbine, water is directed inwards (rather than outwards as in the Fourneyron turbine shown in Figure 2) onto blades attached to the rim of a shaft, which turns an electrical generator. The power is determined by the height of the fall and the volume of water flowing. One of the largest hydropower installations in the world is the Three Gorges Dam on the Yangtze River in China (see Figure 12(b)). The dam has flooded a valley, and created a reservoir of some 1,000 square kilometres in area. The height of water in the dam is about 100 metres above the river downstream, and the flow of around 25,000 cubic metres per second can produce a massive maximum output of 22,500 MW. On average about 90 TWh are generated each year. In China this is sufficient for about 60 million homes. In Europe this electricity would power around 25 million households, while in America about 7.5 million, as energy consumption is higher in these countries.

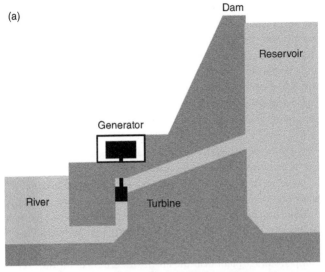

Biomass, solar heat, and hydropower

12. (a) A hydropower plant. (b) The Three Gorges Dam.

Hydropower is now well established, with installations throughout the world that produced about 4,200 TWh of electricity in 2018, some 16 per cent of the global demand. Emissions are very low, and are mainly related to concrete and steel used in construction, and many hydropower plants have operated for over fifty years with low running costs. They are one of the cheapest sources of power and are used for energy-intensive processes, such as aluminium smelting.

A reservoir, made high up in a mountain using a dam, can provide electricity storage. Water is pumped into the reservoir from a river below using the electrical energy to be stored. Power can then be generated when required by letting water out and down through a turbine into the river. Pumped storage plants like this provide over 94 per cent of the world's electricity storage; and more could be provided by adapting some hydropower plants. A seawater pumped storage plant near the north-western tip of the Spencer Gulf in South Australia is being planned, as freshwater resources are scarce. It would have sufficient storage to generate 225 MW for up to eight hours, mitigating pressures on peak demand when temperatures soar.

Environmental and social impact of hydropower

Although hydropower plants can provide large amounts of low-cost, low-carbon electricity, serious social and environmental issues need to be considered when deciding whether the construction of a new hydroelectric scheme is appropriate. These include, in particular, the displacement of population; changes in water quality; the impact on fish; and flooding. In 1954, a devastating flood on the Yangtze River killed 33,000 people, and rendered over a million homeless. The Three Gorges Dam on the river reduced the risk of such disastrous floods, but required 1.3 million people to be relocated. It has been estimated that 30–60 million people worldwide have had to be moved due to hydropower.

Other problems have beset plans to harness one of the largest single hydropower resources in the world; it lies in sub-Saharan Africa, where an estimated 600 million people are without electricity. Near the mouth of the Congo River are the Inga falls, a series of rapids during which the river drops a great height—about 100 metres, which is the height of a thirty-storey building. The river has the second largest flow of water in the world, after the Amazon, and the total potential output is almost twice that of the Three Gorges Dam. This is enough to power enormous industrial growth, and to lift many millions of people out of poverty. But realizing this potential has been severely hampered by wars, corruption, decades of social and political unrest, and massive cost overruns.

Although the risks of a collapse are small with large dams, the consequences can be catastrophic. In 1975, following extreme rainfall, the Banqiao dam in China failed, causing many deaths. There have also been many dams that have been more expensive than budgeted for. Smaller projects, such as run-of-river schemes, in which advantage is taken of a natural fall in water level to avoid a large dam and reservoir, are cheaper. The lack of a large reservoir means that their output is more susceptible to changes in rainfall, but their impact tends to be less disruptive for the environment. Climate change is already causing droughts and larger variations in rainfall in some regions, and is giving concern in countries reliant on hydropower.

For small or isolated communities with a local supply of flowing water, micro-hydropower installations (typically 5–100 kW) can provide an economical source of electricity, and enable them to be self-sustaining. In the developing world micro-hydro installations abound, for example, in remote communities in the Andes and Himalayas, and in hilly parts of the Philippines, Sri Lanka, and China. Micro-hydro schemes such as the Garman water turbine can be locally manufactured, and can also complement solar power, since river flows are typically highest in the winter when solar insolation is lowest.

The potential for hydropower

The resource for hydropower in Europe is quite well exploited, but the potential for growth elsewhere is considerable, particularly in Asia, North and South America, and Africa, and will be helped in some of these regions by low labour costs. About 25 per cent of the global potential of around 15,000 TWh had been realized by 2015 and the International Energy Agency estimates that electricity from hydropower will nearly double by 2050, mainly in developing countries, to about 7,000 TWh per year—about 15 per cent of the expected global electricity demand. As they can be turned on or off when required, hydropower plants can help compensate for the variations in output from wind and solar farms; as will the anticipated increase in the number of pumped storage facilities.

Mountainous countries like Norway and Iceland generate a large fraction of their electricity from hydropower; as do Brazil and Paraguay, which share the 14 gigawatt (1 GW = 1,000 MW) output from the world's second largest hydroelectric power station on the Itaipu dam. This dam is on the Parana river, which is located on the border between the two countries. However, in countries where the resource is less abundant, hydropower is mainly used to satisfy peak demand. The output of many hydropower systems could be raised by about 20 per cent through renovation and modernization, which may be more cost effective and socially acceptable than large new projects. Research on minimizing the environmental and social impact of hydropower is particularly important. But these concerns must be balanced against the need to provide energy to raise the standard of living.

Chapter 4
Wind power

The origin of wind is energy from the Sun. Solar radiation is absorbed mainly by the land and the sea, and in turn they heat the surrounding air. The warm air rises, and when the heating is uneven this movement can create a wind. By a coast, the sea retains heat from the Sun better than the land. So, at night after a sunny day, the rising warm air over the sea draws in cooler air from over the land, and generates a wind that blows out to sea (see Figure 13). On a global scale, the high solar intensity at the equator causes warm air to rise and cooler air to flow in from the north and south. The rotation of the Earth means that in the northern hemisphere the prevailing winds come from the north-east; in the southern hemisphere these winds come from the south-east: these trade winds occur within the 30° latitudes. Above these latitudes, which approximately enclose Africa, the large-scale movement of air gives rise to the Westerlies. The actual distribution of winds is significantly affected by the locations of the continents and oceans, and by the seasons. The winds are strong and reliable in many regions, and it is in these that the energy in the wind can best be harnessed.

The exploitation of wind resources for generating electricity really accelerated when oil prices jumped in the 1970s. Initially, many different designs of modern wind turbine with both horizontal and vertical axes were explored. Those with a vertical axis, looking

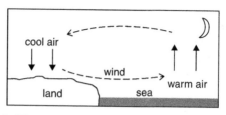

13. Coastal offshore wind at night.

often like a vertical egg whisk, can have their generators closer to the ground, which makes maintenance easier, but then lose the advantage of the stronger winds that occur at greater heights. For large machines, their cost-effectiveness was not as good as those with a horizontal axis: their blade shapes were less efficient, and they tended to be heavier and as a result costlier. Horizontal axis machines are now the dominant design for large power outputs.

Modern wind turbines

A typical modern wind turbine has three blades mounted on a horizontal wind shaft, each some 60 metres long, which drive an electricity generator mounted on top of a tower. The tower is about 100 metres tall, more or less the same height as a thirty-storey building, so a turbine is an enormous machine. Each blade has an aerofoil cross-section and, just as air flowing over an aircraft wing generates lift, the turbine blades experience a force from the wind passing over them that makes them rotate. The wind loses speed as it pushes the blades round, as power is extracted, which makes the wind slower downwind of the turbine. The blades are sufficiently wide to extract a good fraction of the wind's power, but not so large that they act as a significant obstruction; as then most of the wind would be diverted around the turbine by the pressure of the air in front of the blades, rather than push the blades round. Having three blades on a turbine gives a balanced rotation and good performance—one would be unbalanced, while two give a disconcerting flickering effect as the blades pass the tower, and the marginally improved output with four or more blades is not worth the extra cost.

It might appear that the blades on a wind turbine are moving slowly, but while they only rotate at around twenty revolutions per minute, their length means that the speed of the blade tips is very fast—around 125 metres per second, which is 450 kilometres per hour (or 280 miles per hour). At these speeds, the width of a blade near its tip only needs to be about a couple of metres for good power extraction. For a part of the blade closer to the wind shaft, its speed is slower, which is why this section is wider. The blade is also twisted from the tip to the root of the blade to catch the wind at the right angle to maximize the power generated, as the direction of airflow over the blade varies along its length. (At a point on the blade about a tenth of the way out to its tip, the speed of the blade equals the wind speed and the airflow there is at 45 degrees to the wind direction; even further along the blade, the airflow is at a larger angle.)

The blades are oriented into the wind by a motor, and are generally made from composite materials because of their lightness and strength. In early Dutch windmills, the sails were supported by wood, which is a natural composite in which cellulose fibres are embedded in lignin. Siemens have produced 75-metre-long blades for offshore turbines from balsa wood and fibreglass reinforced with epoxy resin. The glass fibres in the fibreglass give it high strength because they are thin enough to be without the flaws that normally weaken glass in bulk. The polymer that encloses them both transfers the loads to the fibres, and also protects them.

Besides their strength, composites have excellent resistance to failure from fatigue. Fatigue involves the initiation and growth of cracks in a material under repeated cycles of stress, at levels that can be considerably less than the material's initial breaking stress. As the blades of a turbine go round, they flex under their weight,

14. An 8 MW wind turbine with 88-metre-long carbon and glass fibre composite blades. This is installed in Bremerhaven, Germany, where it can provide power for about 10,000 homes.

one way and then the other, and they must last about a hundred million revolutions in a lifetime of twenty years, so fatigue resistance is critical. As blades are the highest cost component of a wind turbine, much research and development is directed on better composites; for example, combining glass and carbon fibres in a composite blade can give sufficiently improved performance to more than offset the increase in the cost of the blade (see Figure 14).

Maximum power is generated typically when the turbine is in a wind of around 12 metres per second, which is a 'strong breeze' on the Beaufort scale; windy enough to make holding umbrellas difficult. A turbine with 60-metre-long blades would then generate about four megawatts (4,000 kW), which is called the capacity of the turbine. The power in a wind depends on the cube of its speed, so when the wind speed is halved only one eighth ($\frac{1}{2} \times \frac{1}{2} \times \frac{1}{2}$) of the power is generated, which is why choosing windy sites is so important. Hurricane force winds can have speeds in excess of 70 metres per second and so can have a power some 200 times more than a strong breeze, which is why they can be so devastating. To avoid damaging a turbine in gale force winds, the blades are turned so they do not catch the wind, and brakes are applied to stop any rotation of the blades.

Deployment of wind turbines

Single wind turbines, both large and small, can be used to provide power to homes or a community. A high mounting position is particularly important for small wind turbines (heights of 30 metres or more are best) because, to be economic, the average wind speed needs to be at least 5 metres per second, and the turbulence needs to be small for the blades to be effective. The number of small-scale turbines of less than 50 kW power is increasing, with more than a million estimated worldwide by the end of 2015; around two-thirds of these are in China and a fifth in the USA. They can be deployed in rural locations off-grid to provide electricity or water pumping; and can displace diesel generators; though competition from photovoltaic panels may affect growth in some places.

Wind farms

Wind turbines for large power generation are usually deployed in wind farms, which are arrays of turbines. These are located in regions where the wind conditions are good, such as exposed ridges, high-altitude plains, mountain passes, coastal areas, and out at sea. The turbines are positioned far enough apart so they do not obstruct each other. For 5 MW capacity turbines this means spacing them apart by about one kilometre downwind and two-thirds of a kilometre cross-wind. Tall towers are favourable, as wind speeds pick up with increasing height off the ground (or sea), and can be 30 per cent faster at 100 metres height than at 10 metres. The area of land needed for a 1,000 MW capacity wind farm is about 125 square kilometres, with the land between the turbines still available for grazing or growing crops.

In any location, the speed of the wind is variable, so the output of a wind farm is a fraction of what its capacity would produce. This fraction, the 'capacity factor', is higher for offshore farms, as wind conditions there are generally better than on land, and is typically now about a half, while for onshore farms, capacity factors are approximately a third. A wind farm covering about 30 square kilometres of sea provides 1 TWh of electricity in a year, enough electricity for about 300,000 European homes. On land the corresponding area is about 50 square kilometres.

Offshore wind farms

In countries with a high population density, offshore sites, if available, are more acceptable than those on land, since the turbines are then not so visible. Also, at sea turbines can be much larger, with the blades made close enough to a port that they are not limited in size by the width of roads when transporting them from factory to site (see Figure 15). Capacities of up to 15 MW are planned for offshore wind turbines for the 2020s.

Coastal sea areas have been favoured for offshore wind farms, as the water is shallower and the turbine towers are cheaper

15. Transporting an 88.4 m long blade for an 8 MW offshore turbine.

to construct, and also access to the electrical grid is easier. Impressive advances have been made in foundations for offshore turbines in recent years. Monopiles, which are steel pipes driven about 10 metres or more into the seabed, have been used extensively in the North Sea to support the turbines. In the early 2000s these monopiles were typically 2–4 metres diameter in water depths of 15 metres, and by 2018, 10 metres diameter in depths of 40 metres. The turbines must not obstruct shipping lanes or interfere with radar installations, but even with these restrictions, considerable areas of suitable sites are still available near the coast. All of the UK's electricity demand of 300 TWh per year could be met by offshore wind farms that would only take up about 5 per cent of the area of sea within 50 km of the coast. If the wind farms were located onshore, then 15,000 square kilometres (5,800 square miles) would be needed, which is just over 6 per cent of the area of the UK.

Besides catching the generally better winds farther offshore, wind turbines mounted on floating platforms can be anchored at sea just over the horizon, out of sight of land. These can still be close to demand centres, as some 40 per cent of the world's population lives within 100 km of a coastline. The world's first floating wind farm, Statoil's Hywind farm, is sited in waters 90 to 120 metres deep 25 kilometres offshore off Peterhead in Aberdeenshire in Scotland, and consists of five 6 MW wind turbines that can provide power for over 20,000 homes. It started production in October 2017 and a capacity factor of over 60 per cent has been achieved. Such a high percentage means that power is more likely to be available when demand is high, and helps with integrating the farms' output into a grid. Coupled with this farm is a 1.3 MWh (1,300 kWh) lithium battery store, called Batwind, with a maximum output of 1 MW that can be used to help smooth out the variability in the wind production, adding value to the electricity generated.

The Hywind farm uses a spar buoy platform, developed first for deep water oil extraction. Named after logs that were moored and floated vertically, a spar buoy is a long hollow vertical cylinder with the lower end loaded with ballast, so that the other end is just above water. A spiral fin is attached to the outside of the cylinder to reduce current-induced vibrations, similar to those on tall chimneys—installed for the same reason. The spar's length makes it very resistant to tilting, and the spar buoy is an excellent base for the platform supporting the tower of a wind turbine. The spar buoy design can be used in depths of up to 800 metres, which would open up a huge wind power resource worldwide. In European waters, there is enough wind to meet the total European electricity demand, while off the USA within 200 nautical miles there is the potential to generate about twice the USA demand.

Environmental impact

Wind power produces essentially no global warming or any pollution; only a small amount of associated CO_2 emissions from the fossil fuels used in the construction and operation of the wind farms. And it takes less than a year for a wind farm to generate the same amount of energy used in its manufacture. Noise from the turbines is generally only bothersome when turbines are close to built-up areas. The visual impact of wind farms on the land causes some concern in the UK, though in Germany, which has a similar population density, this is not such an issue. This is probably because more wind farms are owned by communities, rather than commercial companies, and local residents benefit through a share of the annual revenue. In larger, more sparsely populated countries, such as the USA, suitable locations out of sight are more easily found. Besides being more cost effective, it has been found that a few large turbines are more acceptable visually than many small ones.

Turbines do pose a small threat to birds, but in a preliminary study in the USA, the estimated number of bird fatalities per TWh generated by wind turbines was 270, much less than the 5,200 bird fatalities per TWh of electricity from the pollution emitted by fossil-fuel plants. Extrapolating to the UK, about 15,000 birds are killed per year by wind farms; to put this number in perspective, some 55 million birds are estimated to be killed by cats in the UK each year. So, while it is important to locate wind farms away from migratory flight paths or key habitats, wind farms do not pose a significant danger to birds. A much greater long-term threat to birds is from climate change.

The cost of wind power

The global growth in the number of turbines since the 1980s has led to a steady fall in the cost of electricity from wind power. This has come about through economies of scale and technological improvements. In particular, the capacity factors have been increased by using longer blades to improve power generation at lower wind speeds. This fall in costs with the increase in the worldwide capacity is a phenomenon that is seen in a broad range of technologies.

For new technologies such as wind turbines, costs are found to decrease by roughly 20 per cent for every doubling in total capacity; this percentage reduction is called the 'learning rate'. This phenomenon was first identified in the manufacture of aircraft in the 1930s. For onshore wind power, the learning rate was about 19 per cent over the period 1985 to 2014, and saw costs fall from around 57 to 7 cents per kWh as cumulative capacity rose from approximately 1 GW to 350 GW. By 2018, tenders were around 3–5 cents per kWh in several countries. This has meant that onshore wind is now competitive with fossil-fuel generation in many parts of the world, a situation that is termed 'grid-parity'.

Offshore wind has been growing very quickly in the 2010s—in the period 2011–18, the global capacity increased sixfold to 23 GW, with most of the new installations off Europe in the North Sea. Turbines can be taller and larger, giving economies of scale. GE's 12 MW Haliade-X, with blades 107 metres long, will be huge,

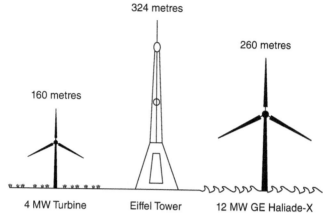

324 metres

260 metres

160 metres

4 MW Turbine   Eiffel Tower   12 MW GE Haliade-X

16. The sizes of modern onshore and offshore wind turbines.

reaching 260 metres above the sea, and will be operational in 2022 (see Figure 16). The costs of construction offshore are greater than on land, because the marine environment can be harsh, and the underwater transmission lines are expensive. But the development and significant deployment of very large turbines is bringing costs down, with new technology, such as 3D printing for prototyping parts, and the use of drones for inspecting wind turbines playing a part. With support for grid connections, electricity from offshore wind farms near coastlines is expected to be as cheap as from fossil-fuel generators by the early 2020s, and from floating wind farms by 2030.

The price of electricity depends on the cost of financing a wind farm. The revenue from the farm must meet the costs of construction, operations, and maintenance. It must also cover the interest that would have been earned by the capital used in the construction over the lifetime of the turbines, typically twenty years. The rate of interest, called the 'discount rate', has a marked effect on the price of electricity; for example, dropping the rate from 12 per cent to 4 per cent could lower the price by almost

40 per cent. Once grid-parity has been reached, and wind power is as cheap as fossil-fuel generation, then interest rates are likely to be less, as the financial risk on the project is reduced, since no subsidies are required. With further deployment both at sea and on land, prices are expected to fall even more.

The output of wind farms is variable and there are costs associated with managing this variability. These depend on the mixture of electricity generators, the interconnectivity with different regions, the amount of energy storage available, and on the ability to vary demand to suit supply.

Global wind potential

What is the maximum energy that wind could generate in a year from wind farms located in all suitable areas on land and in coastal waters? The National Renewable Energy Laboratory in America calculated it would be about 560,000 TWh from those onshore and around 315,000 TWh from those within 200 nautical miles offshore. In these estimates, the wind turbines were envisaged as being spaced to give a capacity of 5 MW per square kilometre. This is fine for the size and spacing of wind farms today, but, if the wind farms covered a substantial fraction of the suitable areas, the spacing would need to be larger in order for the winds to remain high within the wind farms. The exact amount depends on their deployment and is still under investigation, but it could be that the total output is reduced to about 175,000 TWh; however, this would still meet the total global final energy demand of around 100,000 TWh per year.

For wind turbines to produce electricity at a competitive price, the average wind speeds need to be above about 6 metres per second: a moderate breeze that raises dust and moves small branches. Regions around the world on land where the wind is strong enough to install turbines are shown in Figure 17. Many countries have productive winds, and areas with great potential include

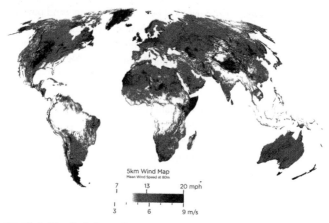

5km Wind Map
Mean Wind Speed at 80m

7   13   20 mph

3   6   9 m/s

17. Global land wind speed map.

northern Europe; parts of the Middle East; west and north China; east and north-west Africa; near the southern tip of South America; and central North America. Estimates of the onshore potentials for China and the United States satisfy each region's electricity requirements very comfortably, although additional resources would be needed to handle wind's variability. Regions where winds are poor include central Africa, the north central area of South America, and parts of South-East Asia.

Outlook for wind power

The global installed capacity of wind turbines was 591 GW in 2018, which generated 4.6 per cent of the world's electricity demand (see Figure 18). In some countries the percentage of demand met by wind power is significantly higher, as in Denmark (43 per cent), and Uruguay (33 per cent). The four countries with the largest capacities are China (207 GW), the USA (97 GW), Germany (53 GW), and India (35 GW). The desire to reduce pollution from coal burning has helped the growth of wind farms in China, but the large wind capacity is overloading the

Renewable Energy

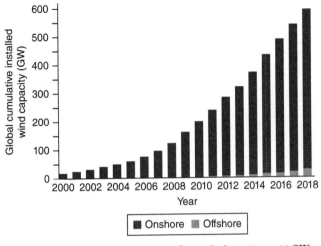

18. **Global wind capacity. Offshore wind capacity in 2018 was 23 GW, and is increasing by around 25% per year. The four countries with the largest offshore capacities in 2018 were: UK 8.3 GW; Germany 6.4 GW; China 4.6 GW; Denmark 1.4 GW.**

transmission grids, and transmission capacity is also affecting growth in India. In Europe, installation of wind farms offshore has been growing fast, and wind farms, both on- and offshore, provided 14 per cent of the EU's electricity demand in 2018. Elsewhere, wind power is expanding quickly.

The global wind power capacity has been growing at an annual rate of 15 per cent since 2010 (see Figure 18). In 2016, the Global Wind Energy Council estimated an average growth rate of 7.5 per cent until 2050, assuming a strong international commitment to meeting climate goals. Then about 5,800 GW would be installed, which would generate around 15,000 TWh per year. This would be a very significant help in reducing our dependence on fossil fuels.

Chapter 5
Solar photovoltaics

In 1839 Edmond Becquerel noticed that uneven illumination by sunlight of electrodes in a solution generated an electric current. But it was not until 1877 that this photovoltaic effect was seen in a solid cell, which was made with the element selenium. A few years later, in 1883, a thin film selenium photocell was produced in the USA by Charles Fritz. But it was expensive, and its efficiency was very low, with only about 1 per cent of the energy from the Sun converted into electricity, so it was not a practical way of generating power. The photovoltaic effect was also seen in other materials, but the photocells that were fabricated from them had similar low efficiencies. It was not until the accidental discovery of a silicon photovoltaic device by Russell Ohl in 1940 in the Bell Laboratories in the USA that significant progress started to be made: the method of making a silicon rod by slow crystallization had caused just the right kind of different impurities to be at either end of a one-inch-long rod for photovoltaic action to be seen, when the rod was illuminated by a flashlight.

Improvements in making silicon for transistors led in 1954 to a cell that had 6 per cent efficiency, about ten times higher than that of earlier devices. This cell, invented by Daryl Chapin, Calvin Fuller, and Gerald Pearson in the Bell Laboratories, was considered to be the first practical solar cell, and there was considerable hype over its discovery. However, the cost of these cells made them practical only in niche applications, such as in satellites. Research led to a lowering in costs by the 1980s, promoting the application of photovoltaic panels for terrestrial use. Programmes encouraging installation of panels on rooftops in the 1990s, and the introduction of feed-in tariffs in the 2000s, generated further demand, and Japan and Germany accelerated the industrialization of production. The average annual growth since 2000 has been almost 40 per cent, and this growth resulted in cost reductions, which in turn helped increase demand.

China responded to the global growth in demand for solar panels in the late 1990s by starting to make panels for the German market. Then later its own domestic market grew, motivated by rising pollution from coal-fired power stations, and concern over energy security and climate change. China concentrated on the dominant technology, crystalline silicon; and government support through loans and tax incentives helped companies build large semi-autonomous factories, which reduced costs dramatically. In the period 2006–11, which saw global production move away from Japan, Germany, and the USA to China, the cost of modules fell by a factor of 3. In the seven years 2011–18, there was a further fall of 3.5 times.

Since their invention, it has taken some sixty years for silicon solar cells' efficiency to increase to over 20 per cent, and for their cost to fall by several hundred times, to the point where the electricity generated by silicon photovoltaic cells can now be cost competitive with that generated by fossil fuels. It has required considerable development and mass production to achieve this, as the processing of silicon to form a solar cell is complex. Silicon cells now account for about 95 per cent of all solar cells. The remainder are based on other photovoltaic materials, such as gallium arsenide and cadmium telluride.

Manufacture of silicon solar cells

For good performance, pure silicon is required as the starting material, and until 1997 the silicon used in solar cells was obtained from the waste from the electronic industry. But with increasing global demand this supply was inadequate. Moreover, the purity required for solar cells is not as stringent as for electronic components, so dedicated plants tailored to solar cell requirements were made. The principal method used is to reduce quartz (silicon dioxide) with carbon in an electric furnace. The silicon is chemically purified and then melted, and a very small quantity of boron added to make what is called p-type silicon.

To produce the wafers for the cells, a monocrystalline p-type silicon rod about 200 mm in diameter and 2–3 metres long is produced by pulling a small silicon crystal out of a crucible of molten silicon. The method was discovered by accident by the chemist Jan Czochralski in 1916 when he accidentally dipped his pen into a crucible of molten tin, rather than into an ink pot, and pulled out a single crystal thread of tin. Alternatively, multicrystalline ingots of silicon can be made by crystallizing the molten silicon; this is a cheaper process, but gives slightly less efficient cells, as there are more defects.

The p-type silicon rod (or ingot) is then cut into thin wafers, which are about 170 microns (2 sheets of paper) thick, using diamond impregnated wires. The front surfaces of the wafers are textured to reduce reflection. Phosphorus is then diffused a very short distance into the wafer to give a very thin surface region of what is known as n-type silicon. Then all but the top thin n-type layer is etched away, and an anti-reflection coating is applied. Finally, contacts are added by screen printing silver paste to the top surface to make fine line electrodes, which minimize shading by only covering a small fraction of the area, and by applying an aluminium layer to the back surface. The surfaces are also treated to optimize charge collection. The challenge in manufacturing cells has been throughout in adapting techniques for obtaining the highest efficiency in a research laboratory into low-cost methods suitable for mass-production.

Operation of silicon solar cells

A silicon solar cell consists of a thick p-type layer, with a very thin n-type layer on top (see Figure 19) where some electrons are free to move about from the phosphorus atoms in the region. These electrons move into the p-type region where they are trapped by the boron atoms. This creates an internal electric field at the junction of the two regions–this p–n junction is crucial to the operation of the cell. (The p and n refer to positive and negative terminals in early experiments.)

When sunlight falls on a silicon cell, some electrons in the p-type region absorb energy from the light and become free to move around within the wafer. When these electrons come within the internal electric field, they are swept into the n-type region. The accumulation of electrons there produces a potential difference that, as in a battery, can drive a current through an appliance (see Figure 19). A minimum energy, 1.1 electron-volts (eV), is required to make the electrons free, and this determines the maximum voltage the cell can produce, which for silicon cells is about 0.7 volts.

In 1905 Albert Einstein realized that light does not only behave as a wave, but also behaves as particles, photons, whose energies are proportional to the frequency of the light. For this insight he received the 1921 Nobel Prize in Physics. Sunlight is composed of light with a range of frequencies, and when it illuminates a silicon solar cell, photons carrying about 25 per cent of the solar energy have insufficient energy to free electrons. Those that do have sufficient transfer all their energy. Any excess energy the electrons have, above that needed to free them, is lost as heat; this accounts

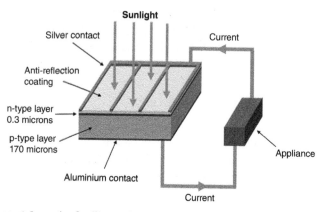

19. Schematic of a silicon solar cell powering an appliance.

for about 30 per cent of the solar energy. A further unavoidable loss of approximately 15 per cent occurs from electrons losing energy through emitting light; adding these losses up gives a maximum theoretical efficiency for a silicon solar cell of about 30 per cent. The best efficiency obtained by 2018 was 26.7 per cent, which is a measure of the enormous development that has been achieved.

The highest efficiency for a silicon solar cell is close to the best that could be achieved with a single p–n junction cell. Photovoltaic materials that require more energy to excite electrons give a higher voltage but a lower current, as only a smaller fraction of the photons in sunlight has sufficient energy to free electrons; while those needing less energy give a higher current but a lower voltage. As the output power is the product of the voltage and the current, there is an optimum value for the 'separation energy' that is close to that of silicon.

Solar cells are typically some 240 square centimetres in area and, since they are very fragile, around sixty are mounted together between glass sheets in an aluminium frame to form a solar module. The cells are connected together in series, giving an operating voltage of about 36 volts. One or more modules make up a solar panel. A solar panel's output is determined by the intensity of the sunlight, and is reduced when it is cloudy to between about a tenth to a third of what it is on a clear sunny day. In the tropics the maximum intensity is close to 1 kilowatt per square metre; typical efficiencies are 21 per cent, so a panel which has a cell area of 1.44 square metres would then produce 300 watts (1,000 × 0.21 × 1.44).

Developments in solar cell technology

One route to improved efficiency is by using both surfaces of a silicon solar cell. These bifacial cells convert light reflected onto the back surface by replacing the aluminium contact there by an aluminium grid, as well as the illumination on the top surface of the cell. By mounting these cells about a metre above a reflective surface such as light-coloured stones or white-painted concrete, the output can be increased by up to 30 per cent, and potentially produce cheaper electricity.

Solar cells can also be made from very thin films of certain photovoltaic materials, and it was thought that these could be much cheaper than silicon cells, as they used much less material. Most types have lost out in the rapid fall in cost of silicon cells, but a few may prove very effective. One is the perovskite solar cell, whose efficiency improved dramatically from less than 5 per cent in 2009 to over 24 per cent by 2019. Perovskites are crystals with the same structure as the mineral calcium titanium oxide, and excellent cells have been made with the perovskite methylammonium lead trihalide. It absorbs photons with energies greater than 1.6 electron-volts (eV) and needs to be only a third of a micron thick—very much thinner than a silicon cell. The absorption energy can be changed by altering the composition of the perovskite. The thin film of perovskite can be simply applied, which should enable production costs to be low, either on solid or flexible surfaces, and there are already plans to make the cells commercially available.

They can also be used to improve the efficiency of a silicon solar cell, and this may be a particularly cost-effective application. This can be done by applying a layer of perovskite on top of the silicon cell. In such a double p–n junction (tandem) cell, photons with energies between 1.1 eV and 1.6 eV are absorbed in the silicon, while those with energies greater than 1.6 eV are absorbed in the perovskite. The tandem cell acts like two batteries in series, with the upper one delivering current at a voltage of 1.2 volts, the lower at 0.7 volts. The current is about half that from a single silicon cell but the voltage at 1.9 volts is more than twice this, so the efficiency could increase from around 22 per cent to 30 per cent. In December 2018, a perovskite–silicon tandem solar cell, manufactured by the company Oxford PV, achieved an efficiency of 28 per cent—higher than the record single junction silicon efficiency of 26.7 per cent. As the gain in efficiency is expected to be much higher than the increase in cost of adding the perovskite layer, the cost of electricity from these tandem cells could be significantly lower than from silicon cells. The company, together with the University of Oxford, is also exploring the possibility of making a low-cost triple junction solar cell, with three different perovskite layers that absorb photons of differing energies, which could potentially have 37 per cent efficiency.

Another area in which progress is being made is in fabricating organic thin-film solar cells on flexible plastic substrates. If printing technology can be used, then vast areas of solar cells might be produced quickly and cheaply (compare the printing of newspapers). These cells can have a coloured appearance, which can be attractive on buildings. At present (2019), single junction cells have efficiencies over 16 per cent, and tandem ones 17.3 per cent. Modules weigh around 0.5 kg compared with about 11 kg per square metre for silicon panels. With economies of scale these may prove a valuable addition, such as for flimsy dwellings in shanty towns.

Environmental impact

Solar photovoltaic power in operation produces no pollutants, no greenhouse gases, and is a safe way of generating electricity. There are no moving parts, which reduces maintenance and also results in no noise pollution, and no water is required (except some for cleaning). In production, some hazardous materials are used, but the quantities are small. With effective safeguards and regulations, the risks in manufacturing solar panels can be kept very small and acceptable. In Europe, it takes between one and two and a half years, dependent on location, to generate the same amount of energy as was used in making the panels. Production mainly uses fossil-fuel energy at the moment; but as panels last thirty years or more, the carbon footprint of their electricity generation is only about 10 per cent that from gas turbine power plants, and this percentage will reduce as more power is generated from renewables. However, most panels are produced now in China, where electricity is still mainly from coal-fired power stations, and these have about twice the carbon footprint of those built in Europe.

In regions with good sunshine, an area of about 15 square kilometres of land is required to generate one TWh per year, which would power around 300,000 European homes (for comparison, about 50 square kilometres of land would be needed for a wind farm to produce this amount of electricity). A photovoltaic farm would cover a good fraction of this area, unlike a wind farm with space between turbines. However, its impact on agricultural production can be reduced by using lower-quality land, such as brownfield sites or desert areas. Its visual impact can also be lessened by integrating photovoltaic panels into buildings, which can save money as smaller quantities of conventional building materials are required: modules are now available looking like roof tiles, and some can be applied to windows.

Solar farms do not have to be built on land—floating installations are also being developed: China has built a 40 MW solar photovoltaic farm on a lake. This has the advantage of not impacting land use, and also of maintaining efficiency as the panels do not overheat. Coastal locations could also be used.

The economics of silicon solar cells

The most significant development in solar cells in the twelve years 2006–18 has been the dramatic fall in their cost by a factor of about eleven (see Figure 20). The cost is taken as that of a module whose area in full sunlight (1 kilowatt per square metre) would generate one watt of power, called one Watt-peak (Wp). The cost has fallen from about $3.5 to $0.3 per Watt-peak, so a module that produces 300 watts in full sun would have cost around $90 in 2018. About 95 per cent of modules are based on silicon.

While efficiencies have increased from 15 per cent to 21 per cent in the last decade, costs have mainly fallen through using thinner

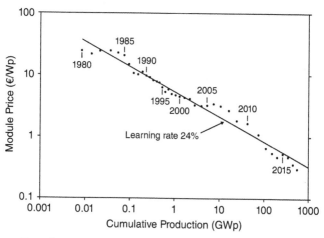

20. Learning curve for solar modules 1980–2018.

wafers, through applying less silver, through developing high-volume processes, and through economies of scale. The learning rate, which is the percentage reduction in costs for every doubling in global capacity, has been 24 per cent, as can be seen in Figure 20. This is in line with Swanson's law, which predicts a percentage drop of 20 per cent. (The bump around 2006 was due to a shortage of silicon.) The cost of a solar panel system depends not only on the cost of the modules but also on the cost of structural and electrical components, and of installation. These balance-of-system costs are now roughly two-thirds for large photovoltaic farms, where costs have been reduced by using robots in assembly; but are more for residential systems, where savings can be made by integrating the panels into buildings.

The cost of electricity depends in particular on that of the solar panel system and on the solar intensity. In the USA, the utility scale cost of electricity has dropped by a factor of around eight during 2009–18: costs fell from about 36 cents to 4.5 cents per kWh, and it is now cheaper than electricity produced from coal, and about the same as from gas; the costs from these fossil-fuel plants are in the range 4–14 cents per kWh. However, the output of photovoltaic farms is variable and there are costs associated with managing this variability. As with wind farms, these depend on the mixture of electricity generators, the interconnectivity with different regions, the amount of energy storage available, and on the ability to vary demand to suit supply. As the amount generated depends on the amount of sunlight, the cost of electricity will generally be more expensive in less sunny regions and countries.

Subsidies are becoming less necessary to promote photovoltaics as their costs have fallen so much, and we are now seeing prices set through auctions. The improvements in technology and the competition that these auctions encourage have seen prices fall in 2018 to as low as 2–3 US cents per kWh in Egypt, India, Saudi Arabia, UAE (Dubai), and the USA (Texas). Prices are also helped

Solar photovoltaics

69

by a low cost of borrowing money, and by favourable support policies. It looks quite possible that by 2030 prices of 2–4 US cents per kWh will be widespread across the globe.

Global solar photovoltaic potential

Figure 21 shows a world map of the annual amount of solar energy per square metre that falls on a fixed surface, inclined at about the angle of latitude to best catch the direct sunlight, and which faces due south or north in the Northern or Southern Hemispheres, respectively. For fixed installations, solar panels are similarly orientated for optimal light collection. Photovoltaics could make a significant contribution to meeting electricity demand in nearly all populated regions.

As can be seen, the variation in sunshine from countries in the tropics to those in Central Europe is about a factor of two. In the south-west of the United States levels of sunlight are such that about 1,700 kWh could be produced annually for every 1 kWp of panels; in Bangladesh and Nigeria 1,400 kWh; in Spain and Japan 1,300 kWh, while in Germany, it is around 925 kWh. These amounts are often expressed as the fraction of what a 1 kWp panel would produce in a year if operating all of the time (i.e. 8,760 kWh), and are then called capacity factors. So, in the US south-west, capacity factors are approximately 19 per cent, in Germany 11 per cent, and these values are reflected in the differing cost of electricity, which was in 2018 around 3 US cents per kWh and 4.5 eurocents per kWh, respectively (€1 ≈ 1.1$).

To estimate the potential contribution that photovoltaics can make to a country's energy demand, we need to evaluate suitable areas. For solar photovoltaic farms, urban areas, forest, ice-covered regions, protected areas, and mountainous terrain have to be excluded. Only a small percentage of agricultural land is suitable,

Yearly sum of global irradiance

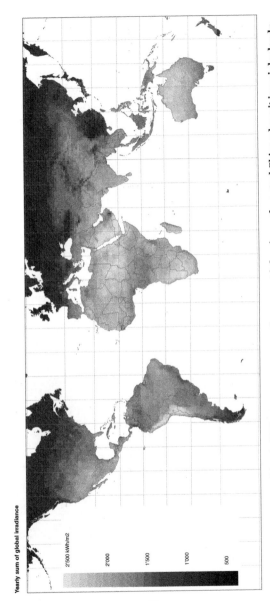

2'500 kWh/m2
2000
1500
1000
500

21. Average annual sunshine in kWh per square metre—the amount is low in south central China where it is mainly cloudy and overcast.

with progressively larger percentages of grassland, barren areas, and deserts. A recent estimate of the potential electricity production from farms covering these areas is 600,000 TWh per year, which is some six times the world's final energy consumption. Furthermore, the amount predicted from panels on rooftops could be considerable: in the USA, the National Renewable Energy Laboratory has estimated 40 per cent of current electricity demand could be met this way.

The growth in solar photovoltaics

Most photovoltaic capacity is being added globally through large farms, such as the one in Qinghai province in China, illustrated in Figure 22, which can have capacities of over 500 MW. (Panels can be mounted on single-axis solar trackers, which increase the output relative to fixed orientation by about 20 per cent, but this is not worthwhile when panels are less expensive than trackers.) A steady expansion is also happening in residential systems,

22. Photovoltaic panels in the Qinghai Golmud Solar Park in China.

where the cost of electricity from domestic rooftop installations is now often comparable to electricity rates from the grid in the USA and in Germany. These systems can give their owners the possibility of selling excess generation back to the grid, called net metering, or storing it in batteries for evening use. About 40 per cent of global capacity growth is in distributed systems, rather than in farms. But there has been concern over net metering as it pushes the burden of maintaining the grid onto those without solar panels.

In the developing parts of Asia and sub-Saharan Africa, it is not just in supplying electricity to grids where solar panels are making such a difference, it is also in distributed generation through rooftop solar installations; and in powering mini-grids in regions where grids are non-existent or poor in quality. Almost one billion people (13 per cent of the world population) are still without access to electricity, mainly in sub-Saharan Africa (600 million) and in India (200 million). In sub-Saharan Africa, the distances are often so great that building a grid would be very expensive, and even where a village is on a grid, the cost of connection for a household can often be unaffordable. The rapidly falling cost of solar panels has meant wider access to affordable clean electricity (see Figure 23). Many homes now have solar power, with modern energy services provided by increasingly efficient and cheap LED lighting and appliances. As battery costs fall, cooking by electricity will be increasingly available, and diesel generators for supply at night will no longer be needed.

Payment for these systems is being made easier for those with little capital by pay-as-you-go (PAYG) schemes coupled with mobile banking. But ensuring finance for these initiatives is not straightforward, and electricity is not yet reaching the very poor. Where both off-grid and grid expansions are taking place, these need to be complementary, and government involvement is important, as, for example, in Nigeria. While the cities and most villages in India are connected to a grid, which provides their

23. Solar panel recharging a mobile phone in Malawi.

schools and public institutions with electricity, there are often many homes in a village that cannot afford or do not want to be connected, as the supply is often unreliable. Solar powered homes and mini-grids can give a more reliable supply in such situations. In Bangladesh, several million households have solar home systems, many on mini-grids that enable electricity trading, using blockchain to ensure secure exchange of information, which helps balance supply and demand.

Outlook for solar photovoltaics

With the cost of producing electricity from utility scale photovoltaic farms now cheaper than fossil-fuel plants in several regions of the world, new generators are increasingly photovoltaic. In 2018, these totalled 100 GW, about the same capacity as added by all non-renewable generators. China dominates, having about a third of the global capacity, and both China and India have ambitious plans for expanding photovoltaic production. But both countries need to upgrade their grids to accommodate their increasing renewable generation. The sharp fall in the cost of solar generation is already helping displace coal generation, with some proposed plants cancelled, but both countries are still building coal plants, which can be easier to bring online to meet local increased demand than a solar farm, unless the farm has storage or other backup supplies. In other countries, an expansion of photovoltaics is also occurring (see Figure 24), and as the cost of

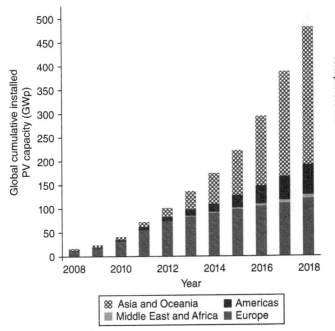

24. Global installed photovoltaic capacity. In 2018, China had 64%, Japan 20%, and India 10% of the Asia and Oceania market, and the USA 79% of the Americas market.

battery storage falls, the investment in solar panels coupled with storage is picking up rapidly.

In 2018 solar photovoltaics provided 2.1 per cent of the world's electricity demand. While this is still small, the global production of solar panels was about 100 GW in 2018, and this is estimated to be doubling every three years. The Fraunhofer Institute estimated in 2015 that there could be a global capacity of up to 15,000 GW generating 20,000 TWh per year by 2050; and with a breakthrough in deployment following massive investment, these quantities might be doubled. Such enormous growth would see the cost of solar photovoltaics fall even more, and would be a way of providing affordable electricity to millions in China, India, North and South America, Africa, and elsewhere.

Chapter 6
Other low-carbon technologies

The traditional renewables (biomass, solar heat, and hydropower) could provide annually by 2050 close to 30,000 TWh, and, with massive investment, wind turbines and solar panels could generate 15,000 TWh and up to 40,000 TWh, respectively. This would bring the total renewable energy supply near 85,000 TWh per year by 2050, approximately 85 per cent of current final energy consumption. What could the other renewables, tidal, wave, geothermal power, and the low-carbon technologies of nuclear power and carbon capture contribute?

Tidal power

Seas have tides because the gravitational pull of the Moon on the oceans on the near side of the Earth is slightly larger, and on the far side slightly smaller, than that necessary to keep the Earth and Moon orbiting each other. As a result, the ocean level is a bit higher on either side (and a bit lower in between), and the weight of raised water compensates for the difference in pull. As the Earth rotates, these bulges cause two tides a day, with the variation in sea level, called the tidal range, about 0.5 m. The pull of the Sun increases or decreases the range by about 20 per cent, depending on its relative position, to give spring and neap tides. However, where a natural oscillation of seawater has a period matching that of the rotation of the Earth, the range can be much larger. This happens for the tides in the Atlantic basin, and also in some inlets and estuaries. If a bay has a funnel shape then that further enhances the range of the tide, as the decreasing width causes an incoming tide to rise up and an outgoing one to drop down.

Locations with large tidal ranges are ideal for tidal power plants. The earliest plants were tide mills, some of which date to before the 10th century. By the 18th century, there were over fifty on the tidal River Thames. Many others were located around the Atlantic shoreline with about 750 in operation at one time. The mills had a storage pond that filled when the tide came in. Around low tide, the water was let out over a water wheel, which typically powered a grinding stone for corn, or a timber saw. These tide mills generated around 10 kilowatts, but enormous powers are possible if an estuary with a high tidal range is blocked to form a huge storage pool.

The first proposal for a large power plant was in 1920 in the Bay of Fundy on the USA–Canada border where the largest tides in the world, around 17 metres in range, are found. A series of dams blocking the Bay of Passamaquoddy would have created two pools separated by a 350 megawatt power station. But the great crash in 1929, and then concern over the cost and its effect on the environment, killed the idea. However, a scheme for a barrage in the La Rance estuary in northern France (see Figure 25), where the tidal range is similar, did come to fruition in 1966 and has produced around 240 megawatts ever since. It was the largest tidal plant in the world until the Sihwa tidal barrage 254 MW plant in South Korea was completed in 2011.

The UK has considerable tidal resources with the largest in the Severn estuary. There the tidal range is about 14 metres and the tidal flow is such that some 8,600 MW could be generated by a barrage. This idea has been discussed since the 1920s and was last considered, but rejected, in 2010 due to concerns over the very

25. La Rance tidal barrage power plant in Northern France.

high cost and the significant effect on the ecology of the Severn. Although the barrage would lower the flow of the river, which would increase biodiversity, it would affect the inter-tidal area, which is an important feeding ground for birds. There was a proposal for a tidal lagoon at Swansea in the Severn estuary, which would have produced about 320 MW with little adverse effect on the environment, since it does not block the whole tidal flow. However, it was also rejected by the UK government, in 2018, as too expensive, particularly when compared with the economics of offshore wind farms in the North Sea. Now schemes to harness the UK potential are focusing on tidal stream projects instead.

Tidal stream power plants are less expensive and potentially more economic than barrages or lagoons. They consist of arrays of underwater rotors located in waters where there are strong tidal currents; the rotors are essentially underwater versions of wind turbines (see Figure 26). Projects are under way in the Minas Passage in the Bay of Fundy, and also off the north of Scotland in the Pentland Firth. In the Firth, the maximum tidal current is

26. An 18 m diameter 1.5 MW tidal stream turbine being lowered into the Pentland Firth in Scotland for the MeyGen project.

5 metres per second and, when turbines are fully installed, around 1,000 MW could be generated. The environment is harsh and the seawater corrosive, and considerable technical challenges arise in cost-effectively manufacturing rotors that can survive. However, the output is predictable and reliable, and the turbines could make a small low-carbon contribution to electricity demand.

Potential for tidal power

The global tidal power resource near coasts is around one million megawatts, but the useful amount is much less, as locations for tidal power are very limited: tidal ranges need to be greater than about 7 metres and currents faster than 2 metres per second for plants to be economic; and only a few per cent (about 500 TWh) of the world's annual electricity demand could be met by tidal power. Considerably more energy lies in ocean marine currents, such as the Gulf Stream, but harnessing these would be very difficult and expensive because of their inaccessibility.

Wave power

Waves are caused mainly by the transfer of energy from a wind to the water surface through friction and air pressure. A typical ocean wave has a wavelength (i.e. the distance between wave crests) of around 100 metres, and an amplitude of 1–1.5 metres. Its power is high at 30–70 kW per metre of wave-front, with most of the energy within a quarter of a wavelength below the surface. The power depends on the square of the amplitude, so a 10-metre wave has 100 times the power of a 1-metre wave. This is a mixed blessing, for although more energy can be extracted, a wave power device has to be very strong to survive extreme waves.

About two million megawatts could potentially be extracted from waves reaching the continental shores, where some 40 per cent of the world's population lives within 100 km of a coastline. This is enough power to supply all their electricity demand, and its potential has attracted many inventors, with the first patent for a device granted in 1799.

However, the difficulties in harnessing the energy in waves because of the harsh conditions at sea, and the availability of cheaper alternative sources of energy, meant there was very little progress in the development of wave power technology until the 1970s. Then the oil price rises prompted much research into renewable energy. Many wave power designs and prototypes were made, but most were discarded as too costly or insufficiently robust to withstand violent storms. A notable 1970s design was the Salter Duck which floated on water and rocked back and forth with the incident waves. It had a cam-shaped cross-section with the circular surface at the rear, so that little of the incident waves would be transmitted by its rocking motion. The Salter Duck had a good efficiency but never progressed beyond small-scale trials in the 1980s, when interest in renewables waned as oil prices fell, and funding was dropped in favour of wind and nuclear power.

Outlook for wave power

In the late 1990s research in wave power technology revived with the increasing evidence for global warming and the need for low-carbon sources, and the volatility of oil and gas prices. The emphasis was now on smaller devices, but designing ones that were cheap enough proved very difficult, with several projects failing to raise sufficient capital to progress. Submerging the wave energy convertor, so that the device can be protected from storms, may be an effective solution. Nearshore, in depths of around 10–20 metres, waves are limited in height because of wave breaking and conditions more stable than far offshore, making nearshore deployment safe from extreme waves, while also reducing the cost of transmission to shore. Carnegie's CETO system off Western Australia will deploy a series of large buoys about 2–3 metres below the surface, each shaped like a flattened sphere some 20 metres in diameter and 5 metres thick (see Figure 27). As a wave passes overhead, the buoys rise up and down, and this motion pumps fluid that drives an electrical generator. It can harvest energy whatever direction the wave is travelling. A similar principle (though with the buoys on the surface and the generators on the sea floor) is used in the Seabased 1 MW array at Sotenas, off Sweden, which started operation in 2016.

Another method that protects the wave energy convertor from storms is to house a generator high above sea level in an open duct that goes down under the surface of the water. A wave causes the water inside the duct to rise and fall forcing air to flow back and forth through a turbine generator. In a Wells turbine, the aerofoil shaped blades are aligned in such a way that it spins the same way irrespective of the direction of air flow. These oscillating water column devices can be incorporated into breakwaters; an example

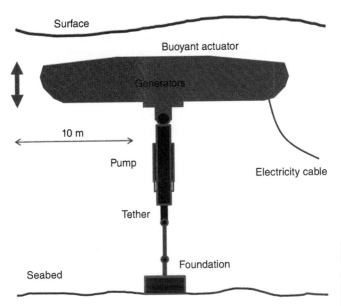

Surface

Buoyant actuator

Generators

10 m

Pump

Electricity cable

Tether

Foundation

Seabed

27. **Carnegie's CETO wave power system.**

being the 300 kW plant at the Mukriku harbour in northern
Spain. Together with submerged devices, which can be deployed
much more widely, these technologies appear to be emerging as
the preferred approach. However, it will be difficult to compete
with the rapidly falling costs of wind and solar photovoltaics, and
production is unlikely to meet more than 2 per cent (500 TWh) of
current global electricity demand.

Geothermal power

The temperature in the interior of the Earth is around 4,000 °C.
It is maintained by the generation of heat produced by radioactive
decay, mainly of uranium and thorium, and by the heat of
crystallization due to the solidification of molten rock. Heat is
conducted through the mantle and within 10 km of the surface

there is a huge volume of hot rock with temperatures up to 300 °C, which represents a vast store of thermal energy. The geothermal resource could sustainably supply about half the world's energy demand, but tapping into this enormous reserve deep underground is very difficult.

Traditionally, geothermal plants have been located near naturally occurring sources, but, as it is difficult to transport heat long distances, the use of geothermal heat is usually restricted to where the source is close to the demand. Geologically active parts of the world such as Iceland, California, Italy, and New Zealand are close to the interfaces between tectonic plates. Here the magma is closer to the surface and the crust can be naturally fractured allowing cold water to seep through to hot rock and escape as high-pressure steam or a mixture of steam and hot water, or as hot springs. These naturally occurring steam jets (geysers) and hot springs provide a ready source of thermal energy—in Iceland nine out of ten houses rely on it.

Geothermal energy is used for district heating, industry, and agricultural purposes, but it can also be used to generate electricity. Figure 28 shows the Krafla geothermal plant in Iceland, which has a generating capacity of about 300 MW and around 130 MW capacity for hot water.

Geothermal power outlook

These geothermal plants provide a small amount of energy, about 0.25 per cent of the global final energy demand at an economically competitive price, and possibly 3 per cent (3,000 TWh per year) by 2050. But much more thermal energy could be 'mined' from below the surface in regions where rock at 250 °C can be found at around 5 km depth, which can be accessed by drilling. For hot rocks the depths must be less than about 10 km, otherwise stopping holes or cracks sealing under the intense pressures and high temperatures is very difficult. The resource potential in the

28. **The Krafla geothermal plant in Iceland.**

USA has been estimated at over 300,000 MW, about a quarter of
the total current generating capacity.

Developments in hot dry rock extraction

The method of extracting the thermal energy is to fracture the
subsurface rock to make it permeable and to then extract the
energy in a similar way to other geothermal sources, by flowing
water through the hot rock. Typically, there will be an injection
well and an extraction well separated horizontally with the region
between them fractured (see Figure 29). The main difficulty is
that rock has a low thermal conductivity. The challenge is in
making a network of fractures that has a sufficiently large surface
area that the temperature of the water rises significantly to give
good thermal efficiency. The fractures must also have narrow gaps,
so that the water flow rate per channel is restricted, as it is the
ratio of heating rate to water flow rate that determines the

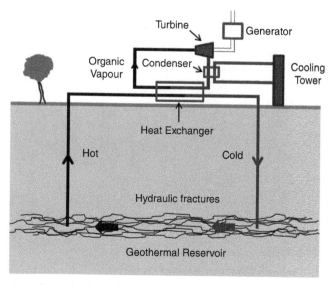

29. Schematic of a hot dry rock geothermal power plant.

temperature rise. Several wide cracks in a network would form a
thermal bypass that would lower the output temperature and
therefore reduce efficiency significantly.

The first hot dry rock plant was in Fenton in the USA at a depth
of 3 km and a temperature of 195 °C and operated during the
1970s–1990s. In that time, it demonstrated the concept with a
power output of a few MW, and research on enhanced geothermal
systems (EGS) expanded. There were pilot plants in Europe, at
Soulz-sous-Forêts in France, in Australia, and elsewhere. Since
2010 at the Newbury site in the USA, AltaRock has created a
network of underground cracks by injecting water at high
pressure. A subsurface array of monitors checked for any seismic
activity to reduce the risk of any serious seismic event, such as
the one in Basel that subsequently resulted in a project being
cancelled in 2009. After forty years, though, hot dry rock

extraction is still under development and costs are currently uncompetitive.

The main expense is in drilling, typically through hard rock, to the depths required, which can be over 60 per cent of the cost of a project. The drilling costs increase almost exponentially with depth, so a breakthrough in drilling technology would make it more cost competitive—using lasers to soften the rock prior to drilling may be effective. It has also proved difficult to fracture the rock reliably, with interference sometimes occurring from natural fissures. A hot dry rock project at Camborne in Cornwall (UK) was eventually abandoned in 1991 as a result of unforeseen problems in the rock formation. However, subsurface evaluation has improved considerably since, and in 2018 attempts at tapping this resource in Cornwall were restarted.

To make the technology more competitive with wind and solar power, research is now focusing on extracting the thermal energy in super-hot (greater than 400 °C) rock at depths greater than 5 km. In 2018, AltaRock signed a partnership with China Coal to evaluate enhanced geothermal's potential to replace coal-fired generation in China. The challenges are significant, but the vast geothermal resource, which is available at all times, makes it important that research and development continues.

Nuclear power

Nuclear power is low carbon as there are no emissions of carbon dioxide from the plant. Commercial power plants are fission reactors, most of which use uranium for fuel. The energy comes from converting part (m) of the mass of the lighter isotope of uranium, ^{235}U, into energy by splitting it into two smaller nuclei, with the energy (E) released given by Einstein's relation $E = mc^2$, which is some tens of million times that in a chemical reaction. This fission is accompanied by the emission of fast neutrons that can initiate another fission that then yields more neutrons; that is, a chain reaction can occur which has to be controlled safely in a nuclear reactor. This is done by balancing the loss of neutrons, which is mainly through capture by the heavier isotope of uranium, ^{238}U, with the gain through neutron-induced fission. The adjustment is made by moving control rods that absorb neutrons in and out of the reactor core.

In a pressurized water reactor, water is heated by circulating it past the uranium-filled fuel rods in the reactor core, where it absorbs the heat generated by the fissions. The water becomes very hot, as it is under pressure, and is used to generate steam to run a turbine generator. The water flowing through the core also slows down the fast neutrons, which increases the probability of neutron-induced fission and reduces the chance of loss through capture. However, to achieve a chain reaction the percentage of the lighter isotope of uranium has to be increased from its naturally occurring abundance of 0.7 per cent; that is, the uranium has to be enriched, typically to a few per cent.

Uranium occurs in many parts of the world, with Kazakhstan, Canada, and Australia currently the main producers. Compared with the amounts of coal, oil, or gas required to fuel a conventional power station, remarkably small amounts of uranium are needed for a nuclear reactor—roughly 1 tonne of uranium will deliver an amount of energy equivalent to 25,000 tonnes of coal. The economic reserves of uranium would give about 100 years of generation at the current level.

Modern reactors

The first reactor was built in a squash court in Chicago in 1942 under the direction of Enrico Fermi. It was part of the Manhattan Project, which developed the first fission bomb, and showed that a nuclear chain reaction could be controlled. Civil nuclear power started with the Calder Hall reactor in England in 1956, followed shortly after in 1957 by the reactor at Shippingport in Pennsylvania, USA. These early prototypes made during the 1950s and 1960s were called generation I reactors. Most of the reactors operating today are generation II designs and were built during the 1970s and 1980s. Global nuclear power output capacity grew quite quickly until the late 1980s, since when it has increased much more slowly.

Its expansion has been affected by the liberalization of electricity markets, which has tended to favour less capital-intensive power plants, and ones with shorter construction times. Three major nuclear accidents have also had a significant effect. In 1979 there was a loss-of-cooling accident in a reactor at Three Mile Island in the USA, resulting in a 20 per cent core meltdown, but fortunately only a small release of radioactivity. Much more serious was an uncontrolled power increase in a reactor in Chernobyl in Ukraine in 1986 that caused a huge release of radioactivity. Although the effect on health was much less than feared, the effect on public confidence was significant. More recently, the Fukushima Daiichi nuclear power plant accident in Japan in 2011 caused renewed concern over the safety of reactors and reduced support for nuclear power, resulting in some countries phasing out their nuclear programmes.

The concern over safety and cost had prompted the development of the generation III reactors in the early 1990s, which were designed to be simpler to build and to include additional passive emergency cooling systems. Passive cooling depends on gravity or temperature differences, rather than on pumps or human intervention, and as a result is expected to be much more reliable and closer to a fail-safe system. A few of these generation III reactors have now been built. However, some recent reactors have proved to be very expensive, due to increased regulatory requirements and lack of nuclear construction experience, and several countries are considering building smaller modular reactors as a way of reducing costs.

Outlook for nuclear power

There were about 450 reactors operating in 2018 with a total capacity of about 400,000 MW, most of them pressurized water reactors, producing about 10 per cent of the world's electricity demand. Nuclear power's share of generation has been falling since 1996, and it is now more expensive in many Western countries than renewables, which are fast becoming cheaper. In Asia, where there is state backing, nuclear is more competitive and some expansion is planned. However, there are still concerns over nuclear proliferation and nuclear waste disposal. No country has yet to build a permanent deep underground site for storing this radioactive material. The International Energy Agency predicts that the global expansion by 2040 will only maintain nuclear power's share of electricity generation at around 10 per cent, with some of the new reactors only replacing those that will be retired. Its contribution could be around 5,000 TWh per year by 2050.

Carbon capture

In the 1990s, when wind and solar power were not developed and concern over climate change was growing, capturing and storing the CO_2 emissions from fossil-fuel power stations was suggested as a way to provide low-carbon electricity. The CO_2 in the flue gases can be chemically isolated and then liquefied by compression and pumped into an underground cavern, such as an aquifer or a disused oil and gas field. At the same time, conventional generators would be required to pay to emit CO_2. This would encourage the adoption of carbon capture technology, provided the carbon price was more than large enough to meet the cost of capturing and storing the CO_2. However, even in the large EU market, the carbon price was never high enough to make carbon capture competitive for electricity production, and very few carbon capture plants are in operation. Even so, capturing

CO_2 emissions could be the most cost-effective way to decarbonize some processes in future. An example is the conversion of natural gas into hydrogen, which can be used for heating and in fuel cells, or in the production of cement and important industrial chemicals, such as methanol and ammonia.

Capturing carbon dioxide from the air is also being seriously considered, as a very real danger exists that CO_2 emissions will not drop fast enough to restrict warming to 1.5 °C. Planting more trees may appear one of the easiest and cheapest ways, but first the huge annual loss of trees must be halted. Already about ten million hectares of forest are being cleared each year for soy, palm oil, and other crops, and also for cattle grazing. This loss causes about a tenth of the annual global CO_2 emissions and a significant loss of biodiversity. Moreover, the area of trees required to sequester a significant amount of CO_2 is huge—about a quarter of the area of the United States for an average each year of 10 per cent of current global emissions from burning fossil fuels, over the six or so decades for the trees to grow to maturity. After this time the trees would need to be replaced, with the timber used in buildings, for example. It has been suggested that the forestry residues could be burnt to produce energy (heat or electricity) and the emitted CO_2 captured and stored. This bioenergy carbon capture is controversial, and care would have to be taken to ensure that the change in land-use resulted in net negative, rather than positive, emissions. Moreover, this method is undeveloped and could be in competition with other demands for arable land and fresh water.

It is possible, though, to capture CO_2 directly from the air using chemical absorbers, which is a much more compact and certain way than using biomass, but it is currently expensive. Origen Power is looking to reduce costs by combining carbon capture with the production of lime, which has a commercial value. Another method, under development by Carbon Engineering, uses potassium hydroxide, which forms potassium carbonate on contact with carbon dioxide. Lime is used to regenerate the potassium hydroxide and form calcium carbonate, which is heated to release the carbon dioxide for compression and storage—in this latter process the lime is regenerated. They estimate that the cost of capturing the CO_2 this way could be as low as $100 per tonne.

Carbon capture outlook

To give added value, the captured CO_2 could be combined with hydrogen (made, for example, by electrolysing water with renewably generated electricity) to make a synthetic low-carbon fuel that could replace gasoline, diesel, or aviation fuel, and have much lower overall emissions than some biofuels. Capturing and storing the carbon dioxide from a coal-fired power station would increase the cost of electricity by around 60 per cent, and using renewables to generate electricity is much cheaper. However, with development and considerable investment, air-capture, together with capture of CO_2 from some industrial processes and through reforestation, might well account for 10 per cent of annual global emissions by 2050.

Total low-carbon potential

By 2050, the total generation from renewables and nuclear power could possibly be close to 90 per cent of current global demand, and with carbon capture the world could have net-zero emissions of CO_2 close to then. But to handle the large amount of renewable electricity, transmission and distribution grids will need to adapt to the variable output from wind and solar farms, and for this the development of energy storage is important.

Chapter 7
Renewable electricity and energy storage

Electricity is vitally important for ensuring a good quality of life. Its generation by wind and solar farms in many parts of the world is increasingly the cheapest production method available, as well as having the great advantage of not emitting any pollution, nor any carbon dioxide that would contribute to global warming. Demand for power is bound to grow with the electrification of transport and of heating across the world. Renewables can give clean and affordable power, and currently supply about a quarter of the global electricity demand. Wind and solar power have the greatest potential, and could supply all the electricity needed, but in 2018 only provided 7 per cent, so a huge expansion is required.

Wind and solar farms are often far from the centres of demand, and so large amounts of electricity must be transmitted over long distances. To avoid any significant heating loss in the transmission cables, the current flowing in them must be low, which means that the voltage has to be very high for the power to be significant. For example, a current of 10 amps requires a voltage of 100 kilovolts to provide 1,000 kilowatts, one megawatt, of power. The voltage and current in a cable are analogous to the pressure and flow of water in a pipe. For a water turbine, it is the flow rate and pressure of water that determines the power; so, to produce much power with a low flow of water requires a very high pressure.

The electric current and voltage in a cable alternate in direction in an alternating current (AC); or remain in one direction in a direct current (DC). Early generators at the end of the 19th century could provide either DC or AC; but it was difficult then, when electricity grids were being erected, to step up DC voltages to the high values needed for efficient transmission. This is why a high voltage alternating current (HVAC) was chosen, because AC voltages could be easily changed using transformers.

The technology has now progressed sufficiently that DC voltages can be changed relatively easily using solid state electronic devices, and high voltage direct current (HVDC) is the more cost-effective method for distances over about 600 kilometres, with losses now of only a few per cent. The main reason is that as the voltage is steady for DC, the voltage can be about twice as high as the effective voltage for AC in an overhead line, before being limited by discharge through the air. Electricity from the Three Gorges Dam is transmitted 940 kilometres to the region of Guangdong in southern China by an HVDC line operating at 500 kilovolts and rated at 3,000 MW (million watts). Also, HVDC is preferred for underground cables, where it has lower electrical losses than HVAC, which is why, for example, the electrical link under the English Channel is by HVDC.

Electricity grids

Although in the 19th century electricity was generated close to where it was needed, economies of scale in the 20th century led to centralized power plants, long-distance power lines, and local substations. Now in most countries in the world electricity is supplied by such a grid. This system is designed for the supply to meet the demand, with the minimum demand, called the baseload, met by the cheapest generators. Up until recently, these were typically coal fired (or nuclear or hydropower plants, where available) and ran for most of the time. They were supplemented by other power stations, usually combined cycle gas turbine plants, to meet the daily variations in load, and by fast-acting gas turbine or diesel generators for surges in demand or for power station outages. Interconnecting transmission lines between the power stations and the substations meant that the electricity supply could be maintained even if a line or power station went down. The grids enabled electricity to be delivered to far-flung communities, as well as accessing remote sources of electricity.

Solar and wind farms are now providing an increasing proportion of electricity on many grids. This is changing the requirements on power plants. Typically, a mix which varies throughout the day using renewable and conventional power plants generates electricity most economically, instead of large conventional generators. In addition to providing clean power, wind and solar farms have the lowest operating costs, called marginal costs, because they have no fuel costs, and are called on first. To ensure that the greatest fraction of generation from wind and solar farms can be accommodated, additional power plants that can respond quickly to changes in supply and demand are best; and preferably these should also run economically at a small fraction of their maximum load. Generally, coal and nuclear plants do not ramp up or down quickly, and gas-fired and renewable plants are better. Depending on location, hydropower, biomass, geothermal, and concentrated solar power (with thermal storage) plants can all be used as flexible generators.

Fossil-fuel power plants can store their fuel and provide electricity on demand. Unlike these generators that can supply at will (called dispatchable or firm supplies), wind and solar farms give a variable output that is dependent on the weather. However, contrary to what some people imagine, grids with significant wind and solar generation are able to provide power when required, despite occasions when there are days with little wind and overcast skies. Variations in the output of solar and wind farms are generally well anticipated, through good weather forecasting that uses artificial intelligence (AI) to obtain the best results.

When the renewable supply is up to 30 per cent of demand, these variations can be readily met with the fast-reacting power plants already installed on the grid to meet changes in demand. Coping with a large 1,000 MW power station unexpectedly tripping (caused by an equipment fault or an overload) can be far more challenging than a sudden drop in wind or solar power. Stand-by reserve plants have to come on line quickly, and wind and solar farms, if not run at full capacity, can provide additional valuable back-up when the weather is windy and sunny, by ramping up their outputs rapidly.

Electricity mainly from renewables

In order to provide clean, secure, and affordable power, and drastically reduce carbon emissions by mid-century to avoid dangerous climate change, we must power grids predominantly by renewables. The percentage of electricity from renewables can be raised to around 50 per cent on a grid by increasing their output, their geographical spread, and by interconnections with other grids. Increasing their generating capacity partly compensates for when weather conditions are poor, and connecting solar and wind farms over a wide area provides a smoother and more reliable output. In Europe, Denmark helps balance supply and demand by trading electricity with Norway, Sweden, or Germany: exporting it when their own wind generation is high and importing it when generation is low.

However, building an intercontinental renewable grid is not straightforward. There was a proposal (DESERTEC) to transmit solar power generated in North Africa to Europe, but it floundered due to political instability, and because there were objections arising from conflicting demands from different regions and countries on the proposed grid. Moreover, the sharp fall in the cost of solar panels has made the advantage of more sunshine less significant: it can be more economic to increase the size of a solar array to compensate for less sunlight, than to pay for the long-distance transmission. Local generation also gives security of supply, by not having to rely on fossil-fuel imports. However, balancing demand and supply can be helped significantly by an extensive grid.

The need for reserve plants can be reduced by altering the demand to match the supply, and this is called demand response. This can be the cheaper option, as the fast-acting power plants used to meet the peak load are the most expensive to run. Adjusting the load to equal the supply can be done using a smart grid that allows two-way communication between the grid operator and the user. This allows just the right amount of demand to be taken off line or added. Many operations exist where interrupting or reducing the electricity supply for a short time is possible: operations with thermal inertia—such as keeping something, such as iron or bitumen, molten, or food in supermarket fridges cold; or when heating or cooling a building—or where a stockpile of items is produced first, before the items are assembled into products. Likewise, demand can be increased by turning on an electric furnace, or a large electrolyser, or (to help cope with droughts from climate change) a desalination plant. We are only at the beginning of the smart grid revolution with its digital technology, which will enable significant changes in load to be made; this will ease the transition to renewable energy and bring lower costs for customers.

Encouraging customers to alter their demand can be done through price differentials. A simple scheme in Italy has the capital for power stations and the cost of distribution recovered through fixed charges that depend on the maximum power used, and production costs through a price per kWh. By restricting the power demand (which makes the electricity cheaper for the consumer), the use of appliances, such as a kettle, a washing machine, and an oven, has to be spaced out during the day; if used all at once the power supply trips. This reduces the peak load for which the cost of generation is highest. Cheaper prices for off-peak (e.g. night time) use is another way. But for better adjustment a smart grid and smart meters are required. Then customers can see details about their consumption, and opt to only use certain devices when electricity prices are low or have an override high price button.

High percentages of renewable generation are greatly helped by having electrical storage available. A mix of solar and wind farms with the capacity to meet the demand in an evening will tend to over-produce during the day, and cause electricity prices to fall. Without storage, this surplus must be exported if possible, or lost through curtailing the supply. Short-term storage can shift some production from the afternoon to the evening, so a smaller capacity can meet the daily demand. As the cost of batteries falls sharply, this storage is increasingly becoming available, and is also starting to displace fast-reacting fossil-fuel plants.

The storage can be either by the generator or by the consumer, and how storage, interconnectors, and a smart grid can help balance supply and demand is illustrated schematically in Figure 30.

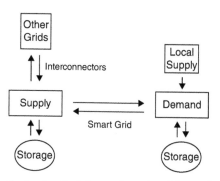

30. Schematic showing the balance of supply and demand on a modern grid.

The cost of changing to renewable generation comes not only from the huge investment in wind and solar farms, but from the cost of strengthening the grid to transport electricity from remote sites. In Germany, the rate of growth of variable renewable generation was slowed while transmission lines were upgraded; and in China rooftop solar panel installations have been encouraged, which will lessen the need for long-distance transmission and reduce the load on the grid. In some countries, local generation with solar panels can be the best source of electricity, particularly where no grid exists, as in much of sub-Saharan Africa; where grid electricity is very expensive, as in parts of Australia; or where the grid has been poor, as in India.

While the cost of producing electricity from wind and solar farms is now very competitive with fossil-fuel generation, the variability of solar and wind adds additional costs to operating a grid. This is because of the expense of back-up generators, for when the supply is low, which adds to the consumer's electricity price. But with greater percentages of renewable generation, fewer large conventional plants are required; and as wind and solar farms are becoming cheaper, this helps offset the cost of the additional reserve plants that must run occasionally to balance supply and demand.

Large amounts of wind and solar generation can give surpluses that lower the farms' revenues, and also those of the reserve plants that operate for less time. (This has been aptly called 'the missing money problem'.) Electricity power markets, introduced in many countries to promote competition in the 1980s and 1990s, are having to adapt to support short-term generation. But interconnectors between different networks, like those connecting grids in the USA or in Europe, can utilize excess production and provide more certain supply. Demand response and storage also lessen the need for back-up plants. These actions increase the value of electricity from solar and wind farms, and, together with their falling costs and cheaper batteries, will help make electricity prices lower.

In order to promote renewable generation in the 1990s and 2000s, when the price of wind and solar power was uncompetitive, subsidies were introduced. The cost of these was typically spread across all users' electricity bills, raising the price of electricity a relatively small amount. This support led to a huge increase in global production of wind and solar generators and a large fall in their cost. In some countries, this has led to a 'boom and bust' cycle of investment in renewables, as subsidies have been given and then cut. Many of these will be paid off for existing farms by the end of the 2030s, and new wind and solar farms are increasingly being built without subsidies. These farms are now not only competitive with fossil-fuel plants, they are fast becoming the cheapest option.

However, a fair comparison needs to consider the considerable subsidies that fossil fuels attract, and in the case of nuclear plants, the liability insurance costs that are picked up by governments. In addition, costs from pollution and from global warming caused by fossil-fuel power plants can be considerable. In many countries these so-called 'externalities' are starting to be addressed by adding a carbon price to all sources of CO_2 emissions, but generally the carbon price is not high enough to reflect the true cost of the damage caused.

With demand response, distributed generation, interconnectors, and short-term storage, the fraction of renewable generation on a grid can be very high. A study for Europe of the costs of increasing the amount of variable renewable energy on the grid from 20 per cent to 80 per cent found prices per kWh rising from around 5 to 8.5 eurocents, with most of the electricity generated by wind power; with low cost for both solar power and batteries, the range went down to about 4 to 6 eurocents. Variations in supply would be principally handled by importing and exporting electricity, together with flexible generators (gas powered). The increase in cost would be far less than that from the damage caused by continuing to burn fossil fuels.

Emissions can be further lowered when there is access to biomass or hydrogen-fuelled turbine generators, concentrated solar power plants with storage, or flexible nuclear power or hydropower plants, which can reduce the dependence on natural gas-fired generators to balance the grid. Obtaining a high percentage of renewables in some regions would be helped by having long-term (several months) storage to meet the large inter-seasonal mismatch between supply and demand that can occur. Some short-term mismatch may soon be met, as costs fall, by over-capacity solar farms with a day's battery storage. In the USA, a report by the Rocky Mountains Institute in 2018 concluded that advances in renewable energy and distributed energy resources, including batteries, over the last decade meant that these can now provide as reliable a level of supply as new gas plants, at a comparable or lower cost.

Solutions that enable a large fraction of renewable generation to power a grid will depend on the resources available in a region, but one key area where further cost reduction would help significantly is in electricity storage.

Electricity storage

Although most electricity storage is currently provided by pumped hydro plants, which are generally available only in hilly regions (see Figure 31), it is the development of the lithium-ion battery that is starting to have a very significant impact. It is a rechargeable battery that has a high power and energy density that first helped enable the mobile phone revolution, and is now poised to initiate a huge expansion in electric cars, as its performance improves and its cost comes down. Lithium-ion batteries can be used anywhere, and are starting to be used in homes with solar panels, for electricity storage, and for balancing supply and demand on the grid.

Historically, a battery referred to something where several things worked together, as in a battery of cannons. In an electrical

31. The Koepchenwerk pumped-storage plant on the Ruhr in Germany.

battery, it is a series of cells joined together, and the first practical rechargeable battery was the lead-acid battery invented in 1859. This battery consists of a series of lead anodes and lead oxide cathodes immersed in sulphuric acid. On discharge, each anode oxidizes and each cathode reduces to lead sulphate. It can provide large currents and is still used in petrol and diesel cars today, but the lead acid battery's energy and power density are too low to be useful for mobile phones or electric cars.

A major breakthrough in making a lightweight rechargeable battery came in the 1980s and 1990s with the development of the lithium-ion battery. Lithium produces a high voltage in a battery and has a low density, so is an ideal material; however, it is chemically very reactive. But a way of controlling its reactivity was found when it was discovered that lithium ions (Li^+) could be moved into or out of some materials without breaking up their structure—a reversible process called intercalation. In particular, graphite could be used for the anode and lithium cobalt oxide (which has a layered structure like graphite) for the cathode.

102

These discoveries by M. S. Whittingham, J. B. Goodenough, and A. Yoshino (2019 Nobel prize winners) sparked the commercialization of the lithium-ion battery by Sony.

A schematic of a lithium-ion battery is shown in Figure 32. On charging, lithium ions move out of the lithium cobalt oxide cathode, through an electrolyte, and combine with graphite in the anode to form lithium graphite. When a device is attached, a current of electrons flows through it from the anode to the cathode, because electrons are more strongly attracted to the lithium cobalt oxide than to the lithium graphite. The current is driven by a voltage of about 3.7 volts and powers the device. During discharge, lithium ions are released into the electrolyte at the anode, and electrons into the external circuit. While at the cathode, the electrons from the external circuit combine with lithium ions, which originate from the anode, and cobalt oxide to form lithium cobalt oxide.

The electrical energy per kilogram of a lithium-ion battery is about a quarter of a kilowatt-hour per kilogram, which is some ten times higher than that of a lead acid battery. It is now the battery of choice for electric vehicles, and increasingly also for short-term storage. Lithium-ion batteries now dominate the market, and as global production rises, their costs are falling fast—some 18 per cent each time the amount manufactured worldwide doubles. They are predicted to cost about $100 per kWh in 2024, and by around then electric vehicles will be cost competitive with conventional cars.

Vehicle-to-grid

An added advantage is that the batteries in electric cars could be connected to the grid, when the cars are parked, and could provide very large amounts of storage for balancing supply and demand. A study for California showed that several billion dollars could be saved if electric vehicles were used in place of stationary storage. But for long-term grid storage (several days or longer),

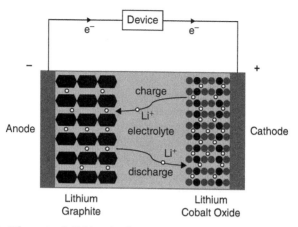

Anode

Device

e⁻

e⁻

−

+

charge

Li^+

electrolyte

Li^+

discharge

Cathode

Lithium
Graphite

Lithium
Cobalt Oxide

32. **Schematic of a lithium-ion battery.**

flow batteries may be more economical than plate batteries, and provide a breakthrough in large-scale storage.

Flow batteries

Flow batteries store their electrical energy within their electrolytes rather than within their electrodes, so their capacity is limited only by the volume of their electrolyte containers—the power and capacity of the battery are therefore decoupled. They can have a high efficiency over a very large number of charge and discharge cycles. Typically, the amount of energy stored per unit mass is between 10 and 50 Watt-hour per kilogram, which is small compared with the amount that can be achieved with plate batteries (about 250 Watt-hour per kilogram for a lithium-ion battery), but they could scale up more cost-effectively, and their capacities can be very large.

The most developed and already commercialized example is the vanadium flow battery. Vanadium is a silvery-grey metal, trace amounts of which considerably increase the strength of steel.

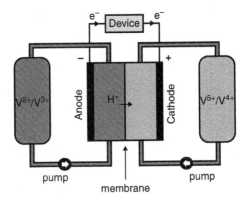

33. Schematic of a vanadium flow battery.

In the time of the Crusades, this element may have contributed to the fabled performance of Damascus swords. Its use in a flow battery takes advantage of its ability to be in several charge states. The battery has two separated uncharged electrolytes, each containing vanadium ions in different positive charge states, plus negatively charged sulphate ions in sulphuric acid. A schematic of this flow battery is shown in Figure 33. As the electrolytes are pumped past the electrodes, vanadium V^{2+} ions are oxidized to V^{3+} ions at the anode with the release of an electron, and V^{5+} ions are reduced to V^{4+} ions at the cathode by the addition of an electron. The electrolytes are separated by a membrane, which allows H^+ ions to pass through to maintain charge neutrality; and the electrodes have catalysts on their surface to speed up the reactions.

However, vanadium is an expensive material and several flow batteries with different chemistries are now under development. As the energy storage requirements on the grid are many millions of kWh, millions of tonnes of material will be needed, so cheaper materials are required. One promising candidate is a sulphur flow battery being developed by Form Energy and supported by

Breakthrough Energy Ventures, a billion-dollar fund chaired by
Bill Gates to fund new energy technologies.

Power-to-gas and other storage technologies

Pumped hydropower accounts for over 94 per cent of global
electricity storage. The regions where this type of storage can be
used could be extended by pumping water between underground,
or underwater, and surface reservoirs, but its deployment is still
likely to be limited. Another technology capable of storing large
amounts of energy utilizes compressed air. In this, electricity is
used to pump air into a huge cavern, where it is stored at high
pressure. The air is then released through a turbine generator
when electricity is required. But it has been uncompetitive, and
there are only two large-scale plants operational, one in Huntdorf
in Germany and the other in Alabama in the USA. There are
plans, though, to build more efficient plants, with one in the
Netherlands for use with wind farms. And in the UK, it has been
suggested that using porous sandstone under the North Sea for
storing compressed air could provide valuable inter-seasonal
energy storage.

A recent innovation is a variation on pumped storage that could
be used widely. The idea, proposed by the company Energy Vault,
is to raise cheap concrete blocks to store energy, rather than water,
using a six-arm crane. Several thousand 35-tonne blocks would
be stacked to form a tower about the height of a thirty-five-storey
building. When power is required, the blocks are attached to
the crane and dropped, which drives the crane motor in reverse,
generating electricity. The Tata Power company in India have
ordered such a tower system that can store 35 MWh with a 4 MW
peak output and a fast response time.

Another alternative is to use electricity to either heat or freeze
water, with the hot water or ice then stored. Buildings need hot
water at all times; and ice can be used to air-condition them, with

demand for air conditioning expected to increase, particularly in the developing world. Such heating and cooling can be done most efficiently by using heat pumps: typically, 1 kW of electrical energy can provide 3 kW of heating or cooling. Alternatively, the heat transferred to the environment in a thermal power station can be used for heating water in what are called combined heat and power (CHP) plants. In Denmark, over 60 per cent of all heat for homes comes from CHP plants that supply local district heating networks, which include large tanks of water as heat stores. The CHP plants can alter their proportion of electricity to heat to help match variations in the output of their wind farms. Many of these CHP plants burn biomass, and these are important in Denmark's plans to eliminate coal-fired power stations by 2030.

However, heat is expensive to transport and typically distances are restricted to about 30 kilometres. A much more versatile means of storing electricity is to use it to generate a combustible fuel that does not contain carbon, such as hydrogen. This can be transported in pipelines, and then stored and used when required; for example, for heating in industry or households. Hydrogen can be produced by electrolysing water, and, commercially, efficiencies as high as 80 per cent have been reached. When hydrogen is burnt in air, it produces just heat and water vapour, so its combustion does not contribute to global warming. Using it as the fuel for a gas-turbine generator or in a fuel cell could therefore reduce emissions in power generation.

A large fraction of the global energy demand is for heating, nearly all of which is currently provided by burning fossil fuels, while stored heat makes only a very small contribution. Using renewable electricity to directly generate heat, as well as combustible fuels, can provide 'renewable heat', which will be very important in reducing fossil-fuel use and carbon dioxide emissions. These together with electric vehicles will enable the decarbonization of both heating and transport.

Chapter 8
Decarbonizing heat and transport

Much of the discussion on renewable energy focuses on the supply of electricity, which now accounts for some quarter of the world's final energy consumption, and, because of the inefficiency of fossil-fuel power plants, about a third of global energy-related CO_2 emissions. However, another third of the emissions are released from burning coal, gas, and oil for heat in industry and for heating buildings, and a tenth from other processes. The remaining quarter of emitted CO_2 is from using oil-derived fuels in engines that power nearly all of transport.

Using renewable electricity directly to produce heat, as in an electric oven, is currently more expensive than burning gas, or other fossil fuels, to generate heat. So, providing the large heat demand with renewable electricity will be very challenging, particularly for industrial processes that require high temperatures, such as in the manufacture of steel and cement. But doing so will avoid pollution and carbon dioxide emissions. For buildings in many parts of the world, energy is needed to make them comfortable to live in, and about 20 per cent of the world's energy consumption is from space heating and cooling, and from water heating in buildings. The heat required is at moderate temperatures, and a very efficient low-carbon technology to do this that fits in well with renewable electricity is a heat pump.

Heat pumps

An air-source heat pump operates like a refrigerator. It has an electric compressor unit circulating a refrigerant around a loop that goes through two coils, one inside a room, the other outside in the air. The compression and subsequent expansion of the refrigerant inside the unit causes one coil to cool and the other to heat (see Figure 34). When the cold coil is inside, a fan blows air across it to cool the room. This is just like in a refrigerator, where the cold coil is inside the fridge and the hot coil is outside in the air.

With this pump, the amount of heat energy that can be transferred is typically two and half times the electrical energy used by the compressor. Reversing the flow in the compressor unit makes the coil inside the room hotter than the outside one, and the heat pump then heats the room. This is a much more efficient way of heating than using an electric bar (resistance) heater, where the heating is only equal to the electrical energy.

Alternatively, rather than being outside in the air, the source coil can be placed underground, where the temperature below a depth of around 10 metres is relatively constant. These ground-source heat pumps tend to be more efficient than air-source pumps (heat 3–4 times the electrical energy), because the ground is at a more stable and suitable temperature than the surface air temperature, being generally warmer in the winter, and colder in the summer.

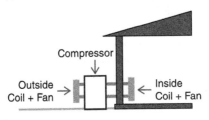

34. Air-source heat pump.

However, their cost is generally higher, as deep holes have to be drilled. Rivers running underground below cities, such as the Tyburn and Fleet under London, could provide good sources for heat pumps in large city buildings. Coils could be placed in these subterranean rivers, which would avoid the cost of drilling boreholes. The technology is already employed in Stuttgart to heat the Baden-Württemberg state ministry using the underground River Nesenbach.

Heat pumps are a very good way to heat or cool buildings but retrofitting them can be expensive, particularly ground-source pumps to existing buildings. Also, the task would be huge: for example, over half of the current buildings in the European Union, Russia, and America will still be in use in 2050, and many of these are not well insulated. Air-source pumps are increasingly being used for new domestic buildings, though pumps using the ground as a source are well suited for new large buildings, where the coils can be installed in the foundations quite easily. By insulating the buildings very well and eliminating draughts, the heating required can be minimal and the electricity for the pumps supplied on site by solar panels.

Clean fuels—bioenergy and hydrogen

A cheaper solution in some countries and locations may be to provide heat by burning a fuel that does not increase carbon dioxide levels in the atmosphere. This can be done with biomass, such as wood pellets, or biofuels, such as biomethane, but suitable supplies of sustainable material are limited. Where they are available, then combined heat and power systems can be used to provide heat and electricity to groups of buildings, as is done in Denmark. Solar thermal panels on roofs can also supplement water heating. But a developing technology that could be widely available and provide a low-carbon substitute for gas is to produce hydrogen, which can be burnt like gas. In France, a demonstration project to blend up to 20 per cent hydrogen in a gas network is under way, while in the UK a pioneering programme to use 100 per cent hydrogen is under consideration.

Over 30 per cent of the emissions in the UK comes from heating and cooking in homes, and this is provided predominantly by natural gas distributed through a gas pipe network. In recent years the piping has been replaced by welded polyethylene pipes. These are suitable for transporting hydrogen as well as natural gas, and the H_{21} project is looking at the feasibility of providing domestic heating using hydrogen, which only emits steam when burnt. Decarbonizing the domestic heating sector would contribute significantly to meeting the UK's Paris Agreement commitments.

Leeds has been chosen for the site of the feasibility study since the city is close to where natural gas is brought on shore and to where the appropriate technology exists for hydrogen production from methane, which is the principal component of natural gas. The city is also near to salt caverns that can be used for storing hydrogen as they seal well. The changeover to hydrogen would be similar to the switch made in the 1970s when town gas was replaced by natural gas, after large deposits were found under the North Sea. (Town gas is mainly hydrogen and carbon monoxide made from reacting steam with coal.)

The plan is to react steam with methane to produce hydrogen and carbon dioxide, a process called steam reforming. The CO_2 would be captured, compressed, and stored under the North Sea in sites similar to the one below the Sleipner gas field in the North Sea, where CO_2 has been successfully stored for many years. Any hydrogen production that was excess to requirements in the summer months would be stored in the nearby large salt caverns, and used when demand is higher in winter.

An alternative way of producing hydrogen is by electrolysing water using renewably generated electricity. Currently this process is more expensive, but its cost is falling. Particular advantages are that it would avoid the cost of capturing the CO_2, and also could be a valuable way of using excess electricity generation from wind and solar farms. In addition, the availability of low-cost hydrogen could help the development of hydrogen-fuelled vehicles.

Appliances (cookers and boilers) would need to be changed to burn hydrogen, but the convenience of using the existing gas network and the ability to store large amounts of energy may make the overall cost of decarbonizing this sector attractive. If adopted, the national network can be converted incrementally with little disruption to customers. The cost could be recuperated over forty years by a small increase in the bills of all customers, which would have little impact on the cost of heating. The hydrogen can be used as safely as natural gas, and could be used not only for heating homes, but for producing high temperature heat for industrial processes.

Heat for industry

It is vital to supply energy to industry with clean fuels, since industry consumes about a third of the global energy demand, and much of it is as heat from burning fossil fuels that vents carbon dioxide and pollution into the atmosphere. Renewably produced electricity can be used directly to supply heat without damaging emissions; such as in electric arc furnaces, which are already employed in steel-from-scrap production, or to produce hydrogen. However, the cost of converting a process to use electricity or to burn hydrogen, and the price of electricity, can make these expensive options, though they may be good choices for new industrial sites, given their long life.

A cheaper solution for producing high temperature heat could be to use electrically heated firebricks, as suggested by researchers at MIT. It is based on a very old technology. Firebricks, baked from a clay that can withstand high temperatures, were used first by the Hittites some 3,000 years ago in their iron smelting kilns. An insulated stack of firebricks could be used as a thermal store, heated up to about 850 °C by electrical resistance heating. For higher temperatures, silicon carbide firebricks might work well. Using excess production from wind and solar farms would be a cheap source of electricity and avoid curtailing the farms' output. When required, air could be blown over the bricks to provide high temperature heat for industrial furnaces at a competitive price, as the materials and air blowers are inexpensive.

Heat pumps are not effective for providing heat at the high temperatures required for industrial processes, and while biomass or biofuels can sometimes be used as a fuel, sustainable supplies are not widely available. Improving the supply chains for forest and agricultural residues and wastes that have little harmful environmental impact (as they do not involve clearing land) would help significantly.

An alternative that is attracting interest is the production of ammonia from renewably generated hydrogen. Ammonia is a valuable industrial product in its own right, being the principal component of fertilizers, but it can also be used in an internal combustion engine, or gas turbine, or fuel cell to provide energy. An advantage is that the infrastructure for transporting ammonia already exists. On combustion it produces nitrogen and water vapour, and some cars have been powered by ammonia, and it may be feasible to power ships this way too. It must be handled carefully as it is toxic at high concentration, but not at the low levels a person can easily detect. Whether this, or other power-to-gas schemes, can contribute significantly to heat supply will depend very much on reducing the cost of production.

Another way to provide heat with low carbon dioxide emissions is to capture the carbon dioxide emitted in existing industrial processes that are powered by fossil fuels. However, the task is made difficult because of the range of processes and large number of sites. A few pilot projects are under way, which can capture up to 90 per cent of emissions. Transport of the carbon dioxide to underground stores, such as depleted oil and gas fields, seems to be quite straightforward and is a small part of the large overall cost. But not all possible sites are near to hand and their long-term security is still under review. Directing efforts on reducing costs for a few major applications, like capturing the carbon dioxide from the manufacture of hydrogen from natural gas, may well be a good strategy. The sharp drop in the cost of renewable energy is making carbon capture increasingly uneconomic for power generation; but with a sufficiently high carbon price it could be a valuable way to decarbonize some processes.

Decarbonizing transport: electric vehicles

Huge carbon dioxide emissions cannot be captured from the millions of internal combustion engines used in vehicles and ships. A massive amount of oil is burnt as petrol and diesel, and accounts for about a quarter of the world's energy use. Renewable substitutes are biofuels, in particular bioethanol and biodiesel. Although there has been a successful programme in Brazil making ethanol from sugar cane, there has been widespread concern elsewhere over the land area required and the competition with food production; and also over the carbon dioxide emissions from land clearances, as in the deforestation in Malaysia from palm oil plantations for biodiesel. One of the most effective ways of reducing these emissions is to avoid combustion altogether by electrifying transport.

The carbon dioxide emissions from passenger cars and vans account for about 11 per cent of the global energy-related CO_2 emissions, and around 15 per cent of the EU emissions. Since 2016 several countries, including France, India, China, Germany, Ireland, the Netherlands, Norway, and the UK, have announced plans to halt the sale of petrol and diesel cars after 2030 or 2040, and emission charges for inner city cars are being introduced.

Across the globe, electric cars are being promoted as a very effective way of reducing emissions when renewables are supplying a significant fraction of the electricity, such as in Norway. Already in the UK, the carbon dioxide emissions per mile are less than half that of the average European conventional car. And these will get even smaller as more renewable generation comes online.

Besides helping to reach the targets for reducing climate change, switching to electric cars cuts out the harmful particulates and nitrous oxides emitted by engines, in particular those powered by diesel. These contribute significantly to pollution in cities around the world, which causes health problems such as asthma. Electric cars emit fewer particulates from their brake pads, as they mainly brake by using their motors as generators. The contribution from tyre wear is similar for all cars; but if cars can be made lighter in future, and acceleration and deceleration reduced by self-driving (autonomous) vehicles, then this will improve.

Cost and outlook for electric vehicles

The performance of electric cars is good, but the main hurdle to switching is their cost. However, new and traditional car manufacturers are already investing a considerable amount of money in developing electric cars. These have many fewer components than petrol and diesel cars, mainly because of the simplicity of electric motors, and have cheaper running costs, but their high initial price is primarily due to the cost of the battery. It is estimated that research and development will bring down battery costs from around $180 per kWh in 2018 to below $100 per kWh by 2024, by which time electric vehicles will be about as cheap to buy as internal combustion engine cars.

Some new electric cars are now equipped with batteries that give a range of 300 kilometres, but to remove the concern about running out of power in the middle of a journey (range anxiety), and avoid costly batteries, there need to be many more charging points, a problem that is starting to be addressed. Batteries typically now take around thirty minutes with a fast charger to be 80 per cent full, which can also be offputting, but faster charging batteries aiming for recharging in under five minutes are being developed. Researchers at Cambridge University have found that replacing graphite with niobium tungsten oxide as the anode in a lithium ion battery may reduce charging times considerably. Developing wireless charging of electric car batteries by burying coils just beneath the road surface is also planned. These coils could top up batteries when the cars are in car parks, at the roadside, or on the road.

Already sales of electric vehicles are growing significantly: hybrid and plug-in hybrid cars will have a share of the market initially, but their complexity makes them expensive, and pure electric cars will quickly dominate as battery prices fall. An estimate by Bank of America Merrill Lynch has predicted electric car sales will achieve 12 per cent, 34 per cent, and 90 per cent of the market by 2025, 2030, and 2050, respectively. (Transitions to new forms of transport have been even quicker in the past: in 1900, 5th Avenue in New York was full of horse-drawn vehicles; by 1913 it was full of motor cars.) Many of these electric cars will be self-driving and shared: improving safety, reducing congestion, and lowering the cost of travel. They will herald a new age of car transport.

But it is not just cars that are being electrified: buses, motorcycles, mopeds, bicycles, and scooters are as well, and their sales are booming around the world. These will help with reducing pollution and congestion in cities, and give affordable transport to many people. However, electrifying trucks, ships, and aeroplanes, which account for nearly half of the energy demand in transport, is much harder, as currently battery storage is often inadequate. Biofuels could provide low-carbon emissions, and some demand might be met sustainably; but providing a large enough supply of biofuels to meet the whole heavy goods demand currently looks very difficult. A valuable alternative could come from fuel cells.

Fuel cells

The fuel cell was invented in 1839 by William Grove but interest in it soon fell away when cheap fossil fuels became widely available. It was not until the 1950s that significant progress was made in their development when NASA, looking for a lightweight power source, decided to use fuel cells to provide electricity for the Gemini and Apollo space capsules, and more recently for the Space Shuttle.

In a simple fuel cell powering a motor (see Figure 35), hydrogen passes through a porous anode, where it is ionized (loses its electron): the positive hydrogen ions pass through a membrane, and the electrons through the motor. The electrons and ions combine on the surface of the cathode with oxygen atoms to make water, which generates some heat. A fuel cell therefore makes electricity and water from hydrogen and oxygen; it is the reverse of the electrolysis of water where electricity passing through water generates hydrogen and oxygen. (Fuel cells produce about an equal amount of heat to electricity, and can be used to provide heat and power in buildings.)

Considerable losses occur in first making hydrogen, storing it, and then converting it back to electricity. The electricity generated is

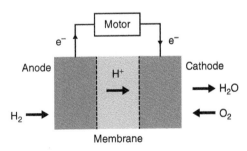

35. Schematic of a hydrogen fuel cell.

only about 35 per cent of the input electrical energy, compared with about 80 per cent for batteries. But the advantages of a fuel cell compared with a battery powered vehicle are the increased range and the much faster refuelling; particularly for trucks, buses, trains, and ships, as these have the space to store the hydrogen. The first hydrogen fuel cell powered trains entered commercial service in Germany in 2018. They are quieter, less expensive to run, and without the polluting emissions of a diesel train, and can be a cheaper alternative to electrifying the railway track. As the global production of fuel cells increases, cost reductions are expected, and their use in place of fossil-fuel generators is likely to increase. With renewably produced hydrogen, fuel cells can provide electricity with no emissions. Hydrogen can be easily transported by pipeline or by truck. Hydrogen's other uses, in particular as an energy store and substitute for natural gas, may well lead to the availability of cheap hydrogen, which will help the expansion of fuel cells.

We can see that providing substitutes for burning fossil fuels for heat or to power vehicles will increase the demand for electricity massively, and this must be provided by renewable energy. Most will come from wind and solar farms, and the challenge will be whether they can be built fast enough, while at the same time transitioning from a world powered mainly by fossil fuels to one dominated by renewables.

Chapter 9
The transition to renewables

The world needs to limit global warming to 1.5 °C to significantly reduce the risks from extreme weather events and the probability of irreversible damage occurring. A 1.1 °C warming has already occurred, so the time for action is short. To keep to this limit, only a further approximately 580 gigatonnes (Gt) of CO_2 can be emitted into the atmosphere (after 2017), which at the current rate will be exceeded by the mid-2030s.

Global carbon dioxide emissions

The total emissions of CO_2 each year depend on the world's annual energy demand, and on the amount of CO_2 emitted per unit of energy produced, which is called the carbon intensity, with coal-fired plants the worst at about one kilogram of CO_2 per kWh. Projections for the total energy demand depend in particular on how the global economy grows in the next few decades. The world's wealth, as measured by the gross domestic product (GDP) of all countries, is increasing at around 3 per cent per year, mostly in the developing world, where it reflects an increase in population and a rise in the standard of living. As a country's GDP grows, the amount of energy it requires increases. This amount of energy required per unit increase in GDP (its energy intensity) reduces over time, particularly through improvements in the efficiency of

Box 1. The Kaya identity showing how carbon dioxide emissions are related to the energy consumption, through the population, personal wealth, energy intensity, and carbon intensity.

$$CO_2 \text{ emissions} = \text{population} \times \frac{GDP}{\text{population}} \times$$
$$\frac{\text{energy}}{GDP} \times \frac{CO_2 \text{ emissions}}{\text{energy}}$$
$$= \text{population} \times \text{personal wealth} \times$$
$$\text{energy intensity} \times \text{carbon intensity}$$

processes, and is currently decreasing worldwide at an average rate of 1.8 per cent. The result is that energy consumption in 2050 is expected to be about 40 per cent higher than it was in 2015. How the global emissions depend on the energy consumption and carbon intensity is summarized concisely in the Kaya identity (see Box 1).

The commitments made by countries in Paris in 2015 to reduce global warming, if all enacted (together with current and planned policies), would mean that only about 70 per cent, rather than 80 per cent, of the global energy supply would be from fossil fuels in 2050, with coal use being much reduced. But this change in fossil-fuel use, combined with the estimated rise in energy demand, would only make the emissions much the same as now, and the cumulative amount of CO_2 emitted in the interval 2017–50 would be more than 1,200 gigatonnes, and would probably take global warming near to 2 °C. What is required is to reduce emissions to zero by about 2050. This can be done by further reducing energy demand and speeding up the shift to renewable energy, but it will require a huge effort across the globe.

Reducing energy demand

Most of the world's energy consumption is in buildings, transport, and industry, in roughly equal proportions, and there is great scope for decreasing energy usage in each of these areas. Already there are passive designs for buildings to consume less energy, responsive to the local environmental conditions, using appropriate materials, such as reflective surfaces in hot climates, to help maintain comfortable temperatures. Good insulation, glazing, and draught reduction are also important, and cooling a house can be achieved by shading, the orientation of the building, and natural ventilation. Any heating that is required can be provided by heat pumps, which are much more efficient than using fossil fuels. Heating only those rooms that are occupied (as is often the custom in Japan), and not those that are empty, also saves energy. It is crucial that new buildings are built to high standards as they last for many decades, but improving the thermal performance of existing buildings can be expensive, and so far, progress has been slow.

In lighting, LEDs have about twice the efficiency of fluorescent tubes, and very much more than incandescent bulbs, and are very long lasting. About 15 per cent of electricity use worldwide is on lighting, and as much could be changed to LEDs, there is considerable scope for using less electricity. Improving the efficiency of appliances also helps—in refrigerators, for example, by using additional thermal insulation and better compressors. In transport, imposing regulations on fuel efficiency for fossil-fuel powered vehicles and speed restrictions can reduce energy demand—lighter less powerful cars could lower fuel consumption by 50 per cent. Electric cars give even greater energy savings, as electric motors have much higher efficiencies than internal combustion engines. Reductions can also come about through urban design and policies that promote public (low-carbon) transport, cycling, electric vehicles, and walking; and by avoiding journeys by using video conferencing and internet shopping. In industry, savings can be made by improvements in heat recovery, the efficiency of processes and motors; and finding alternatives, such as timber structures, to cement and steel that require a lot of energy in their manufacture.

Electrification with renewables brings greater efficiency in several areas, notably in transport and heating. The International Renewable Energy Agency has estimated that this synergy, together with the actions described above, could help achieve a greater fall in energy intensity of around 2.8 per cent per year (as well as a fall in carbon intensity) that could keep the world's energy demand roughly constant at about 100,000 TWh a year.

The target for renewable energy

In 2015, the fraction of energy from renewables was about 18 per cent, mostly from bioenergy and hydropower. This will need to be close to 60 per cent, and the fraction from fossil fuels down to about 35 per cent, by 2037 to be on track for zero emissions by 2050, by which time around 85 per cent of the energy demand should be from renewables, 5 per cent from nuclear, with carbon capture removing the remaining emissions from fossil-fuel use. Figure 36 shows how the contribution of renewables to the global final energy consumption might be in 2037 for the world to be on course. Fossil fuel consumption is limited to oil for transport and gas for heating, with coal no longer used. (If we don't achieve this until 2060, with zero emissions by 2090, then the warming will be limited to about 2 °C, but the consequences of climate change will be significantly worse.)

The growth in electric vehicles, in heat pumps, and in power-to-gas doubles the electricity demand to 50,000 TWh a year by 2037. The higher efficiency of electric motors keeps the transport demand about the same. The use of traditional biomass in the

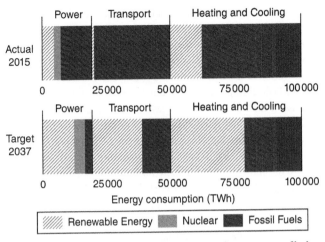

36. **The increase by 2037 in renewable energy to be on target to limit global warming to 1.5 °C by 2050. Power—for machines, appliances, and lighting—was supplied in 2015 by about 80% of the electricity produced; the remaining 20% was used on heating and cooling.**

developing world decreases significantly, with solar panels providing access to electricity for cooking and lighting, but modern biomass for heating and transport has increased its share. Nuclear and hydropower produce some more electricity, but by far the largest increase is in wind and solar power, which will need to supply some 35,000 TWh per year by 2037. But how likely are we to be able to construct sufficient generating capacity in time? And can we reduce our reliance on fossil fuels fast enough?

Reducing fossil fuels

One of the major immediate challenges is to cut the use of coal, as coal combustion causes a third of the world's CO_2 emissions. The capacity of coal-fired power plants almost doubled during the period 2000–18 to 2,000 GW, largely because the Chinese economy grew so quickly. China has now about half the world's total capacity, with the USA and India accounting for another fifth. Coal use also expanded quickly in India, but in both China and India growth is slowing as renewables become cheaper, and because of concern over air pollution and climate change. However, new coal-fired plants associated with China's 'Belt and Road Initiative' are a concern. Coal demand in the USA and Europe is declining (though not quickly in Germany, partly because of their phase-out of nuclear power), and emissions from coal burning look as if they have almost peaked. However, globally they are not falling fast enough to meet the 1.5 °C target: there should be no new coal plants built and existing ones should be phased out more quickly. Oil demand must also fall and this is beginning to happen with the switch from petrol and diesel cars and vans (which account for about 11 per cent of global CO_2 emissions) to electric vehicles; as must the demand for gas, which would be helped by halting fracking.

The speed of the transition to renewables has been greatly hindered by climate deniers in positions of influence, notably in the USA. This denial that climate change is caused by the emissions of carbon dioxide has been promoted by vested interests in the status quo, in particular by the fossil-fuel industry. Many billions of dollars are tied up in companies that own fossil-fuel deposits that need to be left underground, and these 'stranded assets' will become worthless, as will many fossil-fuel power plants. Not only that, but many jobs that are dependent on the combustion of fossil fuels will be lost.

This social cost and upheaval will need to be addressed through financial support and retraining. And the idea of a 'just transition', which is that this cost should be borne by governments rather than the individual fossil-fuel workers, was recognized in the Paris Agreement. Many new jobs are being created by the renewable energy companies worldwide, and the industry will require the investment of many billions of dollars (some of which could come from divesting from fossil-fuel companies). However, the cost of continuing our dependence on fossil fuels would be many times more than that required for renewables, due to the effect on health and the damage to the environment from climate change. These externalities need to be fully appreciated and costed by imposing a carbon price.

Currently a price on carbon emissions is put via several schemes, either through carbon taxes or through emission trading schemes. Taxes based on the amount of carbon dioxide emitted provide an inducement for everyone to reduce emissions, and can apply to transport, as well as domestic and industrial consumers. But the amount required to achieve a certain reduction is uncertain, and can affect the less well-off adversely. Trading schemes, on the other hand, can incorporate an absolute cap on the amount emitted; in these, companies are assigned a number of credits, each of which allows a certain amount of CO_2 to be emitted, that can be bought or sold. In the European Union scheme, over-allocation of credits has led to too low a price (per tonne of CO_2 emitted) for it to put a brake on fossil fuels; as a result, the UK introduced a minimum carbon price that is helping cause the early closure of their coal-fired plants. In the USA, states rather than the federal government are taking the initiative, with schemes already in place in the north-east and in California. China plans to introduce a large trading scheme that should help in lowering their emissions, and in their shift to renewables.

In addition to a price on carbon emissions, the large subsidies for fossil fuels need to be removed. The amount is more than $370 billion a year, compared to $100 billion that renewables receive, even though the G20 countries, which account for 85 per cent of the world's wealth (gross GDP), agreed a decade ago that these should be phased out. Many are consumer subsidies that reduce prices through government price controls on fuels; these are popular policies as they can make, for example, car travel or cooking cheaper. But often only the well off can afford the fuels, and these subsidies divert government funds from helping the poor and from spending on other priorities such as education and health. The global annual health cost alone caused by burning fossil fuels has been estimated as at least six times these subsidies, and shows just how important is speeding up the transition to renewables. Lawsuits have argued, so far unsuccessfully, that fossil-fuel companies should pay for climate-change-related damages.

Increasing renewable energy

There needs to be a very significant increase in both renewably generated heat and electricity. Much of the heat will be generated from electricity, with heat from biomass probably remaining at around 10–15 per cent of the total energy demand, a level that looks as if it would be sustainable. For renewable electricity, the dramatic fall in costs of solar panels and wind turbines in the last decade has meant that electricity from solar and wind farms is now cheaper than new coal or nuclear power plants in many parts of the world; in 2018 such farms provided 50 per cent of new generating capacity.

The growth of wind and solar power has been such that their combined global capacity is now half that of coal-fired power plants, and within a decade is expected to be comparable, as coal generation fades. Recently, the global solar capacity has doubled every three years, and wind every six years. Wind farms offshore are growing even more quickly, and are fast becoming competitive with conventional generation. These growth rates and associated cost reductions suggest that with massive investment global wind power capacity could generate some 10,000 TWh and solar power 25,000 TWh a year by 2037, together some 70 per cent of the target electricity consumption. This will be helped by an increased use of robotics to speed up production, and by the rapidly falling cost of battery storage. The remaining electricity generation could come principally from hydropower, nuclear power, and gas power plants. This would be an aggressive expansion of renewable generation, which, if maintained, would generate some 75,000 TWh of electricity and enable there to be zero emissions by 2050. Such growth is possible, provided there is the political will, and currently some progress is being made.

Huge investment is occurring in China, the European Union, the USA, and India, with China expected to account for over 40 per cent of the increase in renewables in the five-year period 2018–23. There is considerable growth in the USA, despite the negative attitude of President Trump, through the initiatives of states, cities, and businesses. Wind and solar farms are the fastest growing sector, and are putting a strain on transmission grids. Many grids are being upgraded or expanded; for instance, in 2018 four new interconnectors across Europe received funding to aid in the integration of more renewable power. While the deployment of renewables has been slow in Russia, where there is an abundance of gas and oil, some growth has occurred in Japan, where they have reduced their reliance on nuclear power following the Fukushima accident but are still quite dependent on coal. Investment in Latin America, where good hydropower and biomass resources exist, is fast increasing, but Brazil and, in particular, Venezuela have considerable oil resources which may slow their transition to renewable generation. In the developing countries of the world, investment in renewables is growing fast, and distributed generation with solar panels is helping millions gain access to electricity.

Actions required

The transition to predominantly renewable electricity generation, aided by ever improving costs and the increasing role of digital technology, is happening, but not yet fast enough. An even larger fraction of our energy production needs to be low-carbon electricity for transport, heating and cooling our buildings, and in industry. There needs to be increased investment in renewable energy generation, improved building codes and regulations on energy efficiency, and the use of coal and of diesel and petrol cars must be phased out as soon as possible. Decarbonizing heating for industry by electrifying industrial processes is vital and is affordable. It can be aided by using electricity to produce low-carbon fuels, such as hydrogen, together with carbon pricing. Above all, international efforts to increase countries' commitments on lowering emissions to meet the Paris Agreement are crucial. But, despite a UN report in 2019 warning that global CO_2 emissions were still rising, no deal was reached on a carbon price at the climate change conference COP25.

Ways to reduce consumption and the demand for energy are essential as they decrease the rate required for decarbonizing the power supply. Since 1970 the global population has doubled and annual carbon dioxide emissions have increased by two and a half times; the world's wealth (GDP) has also grown by a factor of four. These changes have caused a massive depletion and deterioration in the world's resources. In the last forty years, populations of vertebrates declined by 60 per cent on average, with around one million species of plants and animals now at risk of extinction, and vast areas of forest lost. Also, severe pollution in the oceans and the atmosphere, and a precipitous fall in the number of insects resulting from intensive farming (in particular pesticides) and global warming, threatens a catastrophic loss in biodiversity. Built-in obsolescence, with many products thrown away and not reused, has created massive waste (notably plastics). It is vital that we move away from consumerism to a more sustainable lifestyle, encouraging a circular economy with recycling and reuse. Emissions from land clearances must cease, and significant reforestation must occur. And limiting population, which is helped by the increasing urbanization in the world, family planning programmes, and providing education and empowering women, will lower emissions and the pressure on resources.

Above all, we need to stop the burning of fossil fuels as fast as possible, and effective and fair carbon pricing is needed. This will be exceedingly difficult politically, because the use of fossil fuels is enmeshed in our societies. Continued fossil-fuel extraction is already being justified by relying on carbon capture for a significant fraction of global CO_2 emissions. But this would be very unwise, since the technology for carbon capture is not established at that scale, nor would it generally be as cheap as using more renewables; it may well be only cost effective for selected processes and remove only 10 per cent of global emissions. Schemes designed to reduce global warming, such as through artificially increasing atmospheric aerosols, are also being considered, but much more research is needed to assess whether the benefits of geoengineering outweigh the risks from unintended extreme weather, and they must not distract from curbing CO_2 emissions.

States, cities, and individuals can contribute enormously through promoting the use of renewables, and are already doing so: for instance in California and New York State, San Diego, Jaipur, Hamburg, Toronto, and Bangalore. Electrification of city transport (buses and trains) and of cars and bicycles will speed the transition, as will companies and cities that commit to reducing their carbon footprint. And changing modes of transporting goods, from road and air to rail and shipping, would also help.

Individuals can opt to receive their electricity and gas from renewable energy companies, or generate their own renewable energy through rooftop solar panels. They can also reduce their energy consumption and take public transport, or hire electric vehicles. Community involvement can gain much greater acceptance of wind and solar farms; and movements such as '1 million women' and the 'Sunrise Movement' can inspire people to take action on climate change; as has Greta Thunberg and her advocacy of student strikes for action on climate change. The declaration of the first national climate emergency by the UK parliament in May 2019 followed well-publicized protests by the environmental activist group 'Extinction Rebellion' in the previous fortnight.

For those activities for which no low-carbon solution currently exists, such as flying long distances, and which cannot be avoided, offsetting the carbon emissions—whereby emissions are compensated for by financing a reduction in emissions elsewhere—is a transitional solution. This is best done by paying into an independently certified scheme that captures carbon dioxide, such as through enabling effective and sustainable reforestation, or displaces it through renewable generation. It is not a substitute, though, for making all possible reductions in personal emissions.

Energy policy must urgently facilitate a huge and rapid transition to renewables from fossil fuels, an upheaval akin to going to war. A very ambitious proposal in the USA is for a massive investment in decarbonizing the energy supply within a decade with the jobs created used to tackle inequality. It has been called the 'Green New Deal' from its similarities to the economic stimulus of Roosevelt's 'New Deal' in the 1930s. Any plan must be long term to give results and to give confidence in investment. Pay-back may be slow and may be in conflict with private companies' short-term profits for shareholders, in which case strong regulation or state involvement is required.

Non-CO_2 emissions, for example methane from livestock, can be reduced by using vegetable or cultured protein. Eating less meat would reduce the land area required for food and enable ecosystems to be restored, and help in capturing carbon dioxide through reforestation. And in agriculture, conservation tillage, in which seeds are planted without reploughing the land, can reduce CO_2 emissions significantly. Labelling food and consumables with their carbon footprint would make people aware that some products have high associated emissions, and would help in 'greening' consumption.

> **Box 2.   Key actions to avoid dangerous climate change.**
> - Stop burning fossil fuels—use renewable energy instead
> - Decarbonize and increase the electricity supply by investing in wind and solar farms
> - Electrify transport and heating
> - Reduce energy demand—reuse and recycle
> - Promote the expansion of renewables and associated technologies and infrastructure

Individuals must discuss with family, friends, and colleagues the importance of supporting renewables, reducing fossil fuels, and the need for urgent action. The increase in extreme weather events, such as wildfires and severe flooding, is making people realize that something needs to be done, but governments are not reacting quickly enough. We already have the technology to solve the problem of global warming, but we need to give impetus to its deployment. Renewables are now affordable as the costs of electricity from wind and solar farms have dropped dramatically, in some places halving within the last few years. A huge investment will be required, but nothing like the cost of inaction. And divesting from fossil fuels to renewables is fast becoming the most economic, as well as sustainable, choice. Box 2 summarizes the key actions needed.

The world needs to be powered by electricity primarily from wind and solar farms, and the pace of their expansion is a cause for optimism. The growth of electricity grids, demand response, and cheaper battery storage can keep the lights on, and supply the energy we require. We were once dependent on renewables and need to be so again, so that the world can have power that does not damage our environment, or cause climate change that would profoundly affect our lives.

Further reading

The field is moving so fast that books and articles are soon out of date; particularly about solar and wind power, where significant changes have happened since 2015. Below are some books that are relevant, and references where up to date information can be found.

Aklin, M. and Urpelainen, J. *Renewables: the politics of a global transition* (The MIT Press, 2018)

Andrews, J. and Jelley, N. *Energy Science*, 3rd edn (Oxford University Press, 2017)

Berry, S. *50 Ways to Help the Planet; easy ways to live a sustainable life* (Kyle Books, 2018)

Bloomberg, M. and Pope, C. *Climate of Hope: how cities, businesses, and citizens can save the planet* (St. Martin's Press, 2017)

Goodall, C. *The Switch: how solar, storage, and new tech means cheap power for all* (Profile Books, 2016)

Hawken, P., ed. *Drawdown: the most comprehensive plan ever proposed to reverse global warming* (Penguin, 2017)

IRENA. *Global Energy Transformation: a roadmap to 2050* (International Renewable Energy Agency, 2018): <www.irena.org/publications>

Jelley, N. *Dictionary of Energy Science* (Oxford University Press, 2017)

Klein, K. *This Changes Everything: capitalism vs. the climate* (Penguin Books, 2014)

Klein, K. *On Fire: the burning case for a green new deal* (Allen Lane, 2019)

Kolbert, E. *The Sixth Extinction: an unnatural history* (Bloomsbury, 2014)

Lewis, L. L. and Maslin, M. A. *The Human Planet: how we created the anthropocene* (Pelican Books, 2018)

Marshall, G. *Don't even Think about it: why our brains are wired to ignore climate change* (Bloomsbury, 2014)

Maslin, M. A. *Climate Change: a very short introduction*, 3rd edn (Oxford University Press, 2014)

REN21, *Advancing the Global Renewable Energy Transition* (Renewable Energy Policy Network for the 21st Century, 2018): <http://www.ren21.net/advancing-global-renewable-energy-transition/>

Robinson, M. *Climate Justice: hope, resilience, and the fight for a sustainable future* (Bloomsbury, 2018)

Romm, J. *Climate Change: what everyone needs to know*, 2nd edn (Oxford University Press, 2018)

Sivaram, V. *Taming the Sun: innovations to harness solar energy and power the planet* (The MIT Press, 2018)

Usher, B. *Renewable Energy: a primer for the twenty-first century* (Columbia University Press, 2019)

Wallace-Wells, D. *The Uninhabitable Earth: a story of the future* (Allen Lane, 2019)

Weblinks

Special report on Global Warming of 1.5 °C, International Panel on Climate Change, 2018: <https://research.un.org/en/climate-change/reports>

Affordable and clean energy: <http://www.undp.org/content/undp/en/home/sustainable-development-goals/goal-7-affordable-and-clean-energy.html>

Standard of living, the human development index: <http://hdr.undp.org/en/content/human-development-index-hdi>

UN Climate Change Conferences (CoP) https://unfccc.int/process/bodies/supreme-bodies/conference-of-the-parties-cop

Index

For the benefit of digital users, indexed terms that span two pages (e.g., 52–53) may, on occasion, appear on only one of those pages.

Index

GALAXIES
A Very Short Introduction
John Gribbin

Galaxies are the building blocks of the Universe: standing like islands in space, each is made up of many hundreds of millions of stars in which the chemical elements are made, around which planets form, and where on at least one of those planets intelligent life has emerged. In this *Very Short Introduction*, renowned science writer John Gribbin describes the extraordinary things that astronomers are learning about galaxies, and explains how this can shed light on the origins and structure of the Universe.

www.oup.com/vsi

DESERTS
A Very Short Introduction
Nick Middleton

Deserts make up a third of the planet's land surface, but if you picture a desert, what comes to mind? A wasteland? A drought? A place devoid of all life forms? Deserts are remarkable places. Typified by drought and extremes of temperature, they can be harsh and hostile; but many deserts are also spectacularly beautiful, and on occasion teem with life. Nick Middleton explores how each desert is unique: through fantastic life forms, extraordinary scenery, and ingenious human adaptations. He demonstrates a desert's immense natural beauty, its rich biodiversity, and uncovers a long history of successful human occupation. This *Very Short Introduction* tells you everything you ever wanted to know about these extraordinary places and captures their importance in the working of our planet.

NUCLEAR WEAPONS
A Very Short Introduction
Joseph M. Siracusa

In this *Very Short Introduction*, the history and politics of the bomb are explained: from the technology of nuclear weapons, to the revolutionary implications of the H-bomb, and the politics of nuclear deterrence. The issues are set against a backdrop of the changing international landscape, from the early days of development, through the Cold War, to the present-day controversy of George W. Bush's National Missile Defence, and the threat and role of nuclear weapons in the so-called Age of Terror. Joseph M. Siracusa provides a comprehensive, accessible, and at times chilling overview of the most deadly weapon ever invented.

FORENSIC PSYCHOLOGY
A Very Short Introduction
David Canter

Lie detection, offender profiling, jury selection, insanity in
the law, predicting the risk of re-offending, the minds of serial
killers and many other topics that fill news and fiction are all
aspects of the rapidly developing area of scientific psychology
broadly known as Forensic Psychology. *Forensic Psychology:
A Very Short Introduction* discusses all the aspects of psychology
that are relevant to the legal and criminal process as a whole.
It includes explanations of criminal behaviour and criminality,
including the role of mental disorder in crime, and discusses
how forensic psychology contributes to helping investigate
the crime and catching the perpetrators.

www.oup.com/vsi

RELATIVITY
A Very Short Introduction
Russell Stannard

100 years ago, Einstein's theory of relativity shattered the
world of physics. Our comforting Newtonian ideas of space and
time were replaced by bizarre and counterintuitive conclusions:
if you move at high speed, time slows down, space squashes
up and you get heavier; travel fast enough and you could weigh
as much as a jumbo jet, be squashed thinner than a CD without
feeling a thing - and live for ever. And that was just the Special
Theory. With the General Theory came even stranger ideas
of curved space-time, and changed our understanding of gravity
and the cosmos. This authoritative and entertaining *Very Short
Introduction* makes the theory of relativity accessible and
understandable. Using very little mathematics, Russell Stannard
explains the important concepts of relativity, from E=mc2 to black
holes, and explores the theory's impact on science and on our
understanding of the universe.